The Discourse of Sensibility

STUDIES IN HISTORY AND PHILOSOPHY OF SCIENCE

VOLUME 35

General Editor:

STEPHEN GAUKROGER, *University of Sydney*

Editorial Advisory Board:

RACHEL ANKENY, *University of Adelaide*
PETER ANSTEY, *University of Otago*
STEVEN FRENCH, *University of Leeds*
KOEN VERMEIR, *Katholieke Universiteit, Leuven*
OFER GAL, *University of Sydney*
CLEMENCY MONTELLE, *University of Canterbury*
JOHN SCHUSTER, *Campion College & University of Sydney*
RICHARD YEO, *Griffith University*
NICHOLAS RASMUSSEN, *University of New South Wales*

For further volumes:
http://www.springer.com/series/5671

Henry Martyn Lloyd
Editor

The Discourse of Sensibility

The Knowing Body in the Enlightenment

Editor
Henry Martyn Lloyd
Centre for the History of European Discourses
  and School of History, Philosophy, Religion and Classics
The University of Queensland
Brisbane, QLD, Australia

ISSN 0929-6425
ISBN 978-3-319-02701-2   ISBN 978-3-319-02702-9 (eBook)
DOI 10.1007/978-3-319-02702-9
Springer Cham Heidelberg New York Dordrecht London

Library of Congress Control Number: 2013956511

© Springer International Publishing Switzerland 2013
This work is subject to copyright. All rights are reserved by the Publisher, whether the whole or part of the material is concerned, specifically the rights of translation, reprinting, reuse of illustrations, recitation, broadcasting, reproduction on microfilms or in any other physical way, and transmission or information storage and retrieval, electronic adaptation, computer software, or by similar or dissimilar methodology now known or hereafter developed. Exempted from this legal reservation are brief excerpts in connection with reviews or scholarly analysis or material supplied specifically for the purpose of being entered and executed on a computer system, for exclusive use by the purchaser of the work. Duplication of this publication or parts thereof is permitted only under the provisions of the Copyright Law of the Publisher's location, in its current version, and permission for use must always be obtained from Springer. Permissions for use may be obtained through RightsLink at the Copyright Clearance Center. Violations are liable to prosecution under the respective Copyright Law.
The use of general descriptive names, registered names, trademarks, service marks, etc. in this publication does not imply, even in the absence of a specific statement, that such names are exempt from the relevant protective laws and regulations and therefore free for general use.
While the advice and information in this book are believed to be true and accurate at the date of publication, neither the authors nor the editors nor the publisher can accept any legal responsibility for any errors or omissions that may be made. The publisher makes no warranty, express or implied, with respect to the material contained herein.

Printed on acid-free paper

Springer is part of Springer Science+Business Media (www.springer.com)

To Jo, with love and thanks for your constant generosity and support.

Acknowledgements

I cannot remember this project beginning in any real sense. Rather it slowly emerged as an increasingly coherent project over a long period of time. Its first significant incarnation was as a special stream in the Australasian Society of Continental Philosophy annual conference which was held at the University of Queensland in December 2010. I would like to thank my co-conveners for indulging me as I inserted what was essentially a pet project into a much larger event: Marguerite La Caze, Michelle Boulous Walker, Chad Parkhill, and Andrew Wiltshire. I would next like to thank the participants of that stream and I would like to specifically name those who for diverse reasons are not authors represented in this volume: Romana Byrne, Jonathan Lamb, Jennifer-Jones O'Neill, and Annette Pierdziwol.

My thanks to Stephen Gaukroger.

Thank you to the authors who have contributed to this volume. Particularly I am gratified that so many senior scholars supported a project conceived of by someone only just beginning their career. I thank them for their trust, and for their indulgence in allowing me to 'learn on the job'. Of these authors I would like to thank Alexander Cook and Peter Otto who assisted me with my own writing.

From the beginning, this project has been supported by the Centre for the History of European Discourses at the University of Queensland. The Centre was a major sponsor of the 2010 conference and beyond this has given me a great deal of support in many other ways over many years.

I would particularly like to thank two inspirational scholars. First, Peter Cryle for his many years of judicious advice and constant support. And second, Anne Vila. If I were forced to nominate a point when this project did in fact begin I can think of no better occasion than the moment when Peter handed me Anne's *Enlightenment and Pathology: Sensibility in the Literature and Medicine of Eighteenth-Century France* (Baltimore: Johns Hopkins UP, 1998). He did this because he thought it would be useful to guide my own research and because he wanted to show me interdisciplinary intellectual history at its absolute best. In both these aspects the book continues to inspire me and I am not alone in this; as even a cursory reading of this volume will show much of the research presented herein builds on Anne's text.

Anne was the obvious choice to invite as keynote speaker for the 2010 conference and she was a wonderfully congenial guest. Across the life of this project both she and Peter have continued to be indefatigable resources and guides: I thank them.

Finally, thanks to Kim Hajek for picking me up and carrying me the forty-second kilometre.

Brisbane, Australia, 2013   Henry Martyn Lloyd

Contents

1. The Discourse of Sensibility: The Knowing Body in the Enlightenment 1
   Henry Martyn Lloyd

2. Richard Steele and the Rise of Sentiment's Empire 25
   Bridget Orr

3. Rochester's Libertine Poetry as Philosophical Education 43
   Brandon Chua and Justin Clemens

4. Emotional Sensations and the Moral Imagination in Malebranche 63
   Jordan Taylor

5. Feeling Better: Moral Sense and Sensibility in Enlightenment Thought 85
   Alexander Cook

6. Physician, Heal Thyself! Emotions and the Health of the Learned in Samuel Auguste André David Tissot (1728–1797) and Gerard Nicolaas Heerkens (1726–1801) 105
   Yasmin Haskell

7. *Penseurs Profonds*: Sensibility and the Knowledge-Seeker in Eighteenth-Century France 125
   Anne C. Vila

8. Sensibility as Vital Force or as Property of Matter in Mid-Eighteenth-Century Debates 147
   Charles T. Wolfe

9 *Sensibilité*, Embodied Epistemology,
   and the French Enlightenment 171
   Henry Martyn Lloyd

10 **Sensibility in Ruins: Imagined Realities, Perception
   Machines, and the Problem of Experience in Modernity** 195
   Peter Otto

Contributors

Brandon Chua is Postdoctoral Research Fellow at the University of Queensland node of the ARC Centre of Excellence for the History of Emotions (Europe 1100–1800). He has published articles on Restoration literature and is currently at work on a book manuscript entitled *Politics and the Passions: Rethinking Government and the Heroic Idioms of Restoration Culture*.

Justin Clemens teaches at the University of Melbourne. His most recent books include *Psychoanalysis is an Antiphilosophy* (Edinburgh UP); *The Mundiad* (Hunter Publishers 2013) and, with A.J. Bartlett and Jon Roffe, *Lacan Deleuze Badiou* (Edinburgh UP 2014). He is currently working with Marion Campbell on a book on Milton and Lucretius.

Alexander Cook is an intellectual and cultural historian in the School of History at the Australian National University. His research focuses on the period of the Enlightenment and the French Revolution. He has published in leading journals such as *Intellectual History Review, History Workshop Journal, Criticism* and *Sexualities*. His most recent book is an edited volume entitled *Representing Humanity in the Age of Enlightenment* (Pickering and Chatto, 2013). He is co-editor of *History Australia,* the journal of the Australian Historical Association.

Yasmin Haskell is Cassamarca Foundation Chair in Latin Humanism at the University of Western Australia. She is a Chief Investigator in the Australian Research Council's Centre of Excellence for the History of Emotion: 1100–1800, in which she leads projects on 'History of Jesuit Emotions' and 'Passions for Learning'. Her most recent books are *Prescribing Ovid: The Latin Works and Networks of the Enlightened Dr Heerkens* (London: Bloomsbury, 2013) and (edited) *Diseases of the Imagination and Imaginary Disease in the Early Modern Period* (Turnhout: Brepols, 2012).

Henry Martyn Lloyd is currently a fixed-term lecturer in Philosophy at the University of Queensland. He specialises in the history of French philosophy especially the eighteenth- and twentieth-centuries and has published in the *Intellectual*

History Review, in *Philosophy Today*, and in *Parrhesia: A Journal of Critical Philosophy* where he co-edited a special edition of the journal on 'affect'. His thesis is on the Marquis de Sade context of the philosophy of the Enlightenment.

Bridget Orr is an Associate Professor in the Department of English, Vanderbilt University. She is the author of *Empire on the English Stage* (2001) and many essays on eighteenth-century theatre and voyage literature, New Zealand film, and contemporary Maori writing.

Peter Otto is ARC Research Professor (DORA) at the University of Melbourne. His recent publications include *Multiplying Worlds: Romanticism, Modernity, and the Emergence of Virtual Reality* (OUP 2011). He is currently working on a new selected edition of William Blake's illuminated poetry for Oxford University Press.

Jordan Taylor is a doctoral candidate in the Department of Philosophy at the University of Pennsylvania, and an associate of the Department of Cognitive Science at Macquarie University. His main research focus is the history, philosophy, and psychology of perception and emotion. He takes a particular interest in early-modern accounts of the passions and mind-body interaction. He also works at Penn's Institute for Research in Cognitive Science, where he conducts research on how people understand and interpret scientific explanations (with Deena Weisberg).

Anne C. Vila is Professor of French in the Department of French and Italian at the University of Wisconsin-Madison, USA. She is the author of *Enlightenment and Pathology: Sensibility in the Literature and Medicine of Eighteenth-century France* (1998), as well as many articles on the body in the culture of the Enlightenment. She is currently completing a book entitled *Singular Beings: Passions and Pathologies of the Scholar in France, 1720–1840* and an edited volume on *The Cultural History of the Senses in the Enlightenment*.

Charles T. Wolfe is a Research Fellow in the Department of Philosophy and Moral Sciences and Sarton Centre for History of Science, Ghent University. He works in history and philosophy of the early modern life sciences, and some strands of philosophy of biology, with a particular interest in empiricism, materialism, and vitalism. He has edited volumes including *Monsters and Philosophy* (2005), *The Body as Object and Instrument of Knowledge* (2010, w. O. Gal) and *Vitalism and the Scientific Image in Post-Enlightenment Life-science, 1800–2010* (2013, w. S. Normandin). His current project is a monograph on the conceptual foundations of Enlightenment vitalism.

Chapter 1
The Discourse of Sensibility: The Knowing Body in the Enlightenment

Henry Martyn Lloyd

Abstract This chapter introduces the problematic addressed by this volume by contextualising the object of study, the eighteenth-century's body of sensibility, and the discourse within which this object was constructed. It was in terms of this knowing body that the persona of the eighteenth-century knowledge-seeker was constructed. This chapter has two major purposes. First, in order to situate the individual chapters in their broader intellectual context, it outlines four major components of the discourse of sensibility: vitalist medicine, sensationist epistemology, moral sense theory, and aesthetics, including the novel of sensibility. Second, this essay elaborates those general claims collectively supported by the chapters, drawing together what they contribute to questions of the emergence of the discourse, and key elements at stake within the discourse itself. Four major themes are apparent: First, this collection reconstructs various modes by which the sympathetic subject was construed or scripted, including through the theatre, poetry, literature, and medical and philosophical treaties. It furthermore draws out those techniques of affective pedagogy which were implied by the medicalisation of the knowing body, and highlights the manner in which the body of sensibility was constructed as simultaneously particular and universal. Finally, it illustrates the 'centrifugal forces' which were at play within the discourse, and shows the anxiety which often accompanied these forces.

H.M. Lloyd (✉)
Centre for the History of European Discourses and School of History, Philosophy, Religion and Classics, The University of Queensland, 4072 St Lucia, QLD, Australia
e-mail: m.lloyd@uq.edu.au

1.1 Introduction

Famously—infamously—the Enlightenment thinker is associated with 'reason, truth-telling, and the will to bring about social and political reform', and is not typically associated with feeling or embodiment.[1] It is certainly possible to locate, in this period, both celebrations of pure reason, paradigmatically of course Kant, and enthusiastic supporters of rational or 'enlightened' governance, paradigmatically the *philosophes*. But the Enlightenment is also known for its association with emotion. It is thus equally possible to locate moments which celebrate effusive or lachrymose emotion and which are little concerned with reason: the sentimental novels of the 1770s and 1780s, perhaps. Accordingly, it is possible to speak of both a 'rationalist' Enlightenment, which has been taken up principally by ongoing traditions of philosophy, and a 'sentimentalist' Enlightenment, which has principally been taken up by studies of literature.[2] The Enlightenment of 'reason' and that of 'sentiment' are separated from each other by the structure of the contemporary Academy. Alternatively, they are invoked together in the form of a defining paradox: 'reason and sensibility' becomes a disjunctive conjunction. Caution needs to be exercised here; there are good reasons to suppose that taking these two moments as both paradigmatic and mutually exclusive little represents the way in which the period understood itself. During the period, intellectual pursuits were envisioned as having a distinctly embodied and emotional aspect, and the persona of the knowledge-seeker was considered in terms that drew together mind and matter, thought and feeling.

The essays collected in this volume work to reconstruct that very particular object of eighteenth-century thought, the body of sensibility, and the discourse within which it was constructed. The discourse of sensibility was very broadly deployed across the mid- to late-eighteenth century, particularly in France and Britain, on which national contexts this collection will focus. Sensibility was central to the period's aesthetics, epistemology, medicine, natural sciences, and social and philosophical anthropologies. The Enlightenment's knowing body was the body of sensibility; it was in these terms that the persona of the eighteenth-century knowledge-seeker was constructed.

To invoke the term 'discourse' in this way is to invoke deliberately a broadly Foucauldian framework. As explicated in *The Archaeology of Knowledge*, to engage with the past in terms of its 'discursive formations' is to destabilise the established types by which historians have traditionally navigated, including 'categories, divisions, or groupings', established 'unities' such as the book and the *oeuvre*, or contemporary structures such as 'politics' and 'literature'.[3] Foucault invokes four central features which together can be used to mark the presence of what he calls a

---

[1] Vila, Chap. 7.
[2] Frazer 2010, 1–15.
[3] Foucault 2004, 25–28, 31.

'discursive formation'; two of these are particularly useful for delineating the methodological scope of this collection.

First, and of particular importance for this volume, is the feature that Foucault considered as 'being the most likely and easily proved': that 'statements different in form, and dispersed in time, form a group [i.e. a discourse] if they refer to one and the same object'.[4] More precisely, a discourse can be identified by 'the interplay of rules that make possible the appearance of objects during a given period of time'.[5] Rather than relying on the notion of an already given, singular, or unified object, Foucault's key innovation was that a discourse had the effect of unifying what may otherwise have been taken to be a disparate series of objects. The example Foucault invoked here involved the various characteristics brought together under the 'category of delinquency'.[6] The point was this: a key feature of a discourse was its unifying function, its bringing together of a variety of dispersed historical phenomena to form an object. For this collection, the object in question is the body of sensibility, while the discursive formation which constituted that object is the discourse of sensibility. Accordingly, as I discuss below, a significant focus of the essays collected here is what Foucault would call a 'system of dispersion'[7]: the dispersal, the ambiguities, the 'centrifugal forces' (to invoke Alexander Cook's phrase) which operated within the discourse of sensibility, and which nonetheless all contributed to constructing the body of sensibility.

The second feature of a Foucauldian discourse with particular relevance to this volume is that relating to the formation of concepts.[8] Once again, in question are the unifying and constitutive functions of a discourse, but now the focus is on that series of concepts, otherwise apparently disparate, which it draws together (or creates). For this collection, the primary term is 'sensibility'—it may be defined provisionally as the physiological power of sensation or perception, of sensitivity, and of affective responses—a term which was a central notion from the first half of the eighteenth century, but which was rarely used before then.[9] The term 'sensibility' drew into it several others, including: 'sentimental', 'sentiment', 'sense', 'sensation', and 'sympathy'. These terms will be central to this introductory essay, where they will be discussed in turn in the next section, and to the volume as a whole. They are to be read, as they were used in the period, with a good deal of imprecision; as will become clear, the terms bleed into one another such that they are perhaps best described as a family of concepts, rather than as clearly demarcated individuals.[10]

---

[4] Foucault 2004, 35.
[5] Foucault 2004, 36, also 44–54.
[6] Foucault 2004, 47–49.
[7] Foucault 2004, 41.
[8] Foucault 2004, 66.
[9] Vermeir and Deckard 2012, 7–8.
[10] Vila 1998, 2. See also Festa 2006, 14–15; Cook, Chap. 5.

This collection works against the idea that eighteenth-century sensibility is, or ought to be, the purview of any single scholarly discipline. On the contrary, to inquire into the body of sensibility is necessarily to enter into an interdisciplinary space and so to invite the plurality of methodological approaches which this collection exemplifies. This interdisciplinarity goes beyond merely a diversity of historiographical approaches, it also reflects a feature of the discourse itself; I should stress that the discourse of sensibility, as it existed in the eighteenth century, itself operated at the nexus of diverse historical fields. The novel of sensibility has been the subject of a great deal of attention by literary scholars working in both the British and French contexts. These studies have often noted the interaction of the novel with other genres or disciplines, and have shown that sensibility was not just an aesthetic or literary phenomenon.[11] Anne Vila noted, for example, the place of sensibility in fields as diverse as 'physiology, empiricist philosophy, sociomoral theory, medicine, aesthetics, and literature, all of which were included in the loose confederation of naturalistic discourses then known as the "sciences of man"'.[12] Markman Ellis identified seven fields within which the novel of sensibility operated, including moral sense theory, aesthetics, religion (especially latitudinarianism and the rise of philanthropy), political economy, the history of science, the history of sexuality, and the history of popular culture.[13] For Ann Jessie van Sant, 'The three principal contexts in which *sensibility* was a key idea in the eighteenth century are physiology, epistemology, and psychology'.[14] More broadly, sensibility has increasingly interested historians of science: Jessica Riskin and Peter Hanns Reill have demonstrated at length that the discourse was not confined to aesthetics, nor in scientific terms merely to physiology or natural history, but extended to the hard sciences of physics and chemistry, and accordingly played a significant role in the scientific 'empiricism' of the period.[15]

Under the broad umbrella of contextualist intellectual history, the nine articles collected in this volume draw together the histories of literature and aesthetics, metaphysics and epistemology, moral theory, medicine, and cultural history in order to continue the project of reconstructing the eighteenth-century discourse of sensibility. To situate these individual chapters in their broader context, the first part of this introductory essay outlines four major components of the discourse of sensibility: vitalist medicine, sensationist epistemology, moral sense theory, and aesthetics, including the novel of sensibility. In its second part, this introduction draws together the discrete chapters to elaborate the general claims they collectively support, first in terms of questions of the emergence of the discourse, second in terms of what was at stake within the discourse itself.

---

[11] See Vermeir and Deckard 2012; Packham 2012.
[12] Vila 1998, 1.
[13] Ellis 1996, 8.
[14] van Sant 1993, 1.
[15] Riskin 2002, 7; Reill 2005.

1.2 The Context

Historiography has generally equated vitalism—theories which understand life as sustained by some kind of non-mechanical force or power—with nineteenth-century science, Organicism, and Romanticism.[16] But vitalism was also very much an Enlightenment concern and contemporary scholarship has increasingly recognised the importance especially of medical vitalism for eighteenth-century thought. This has particularly been the case for studies of the French Enlightenment, where the significance of Montpellier Vitalism has been long recognised.[17] In the first instance, it is from Montpellier Vitalism that this collection takes its unifying term '*sensibilité/*sensibility' (although it should be noted that Charles Wolfe in Chap. 8 points out that *sensibilité* is perhaps best translated into contemporary English as 'sensitivity'). The ecstatic definition of the *Encyclopédie* took sensibility to be:

> The faculty of feeling, the principle of sensitivity, or the very feeling of the parts, the basis and conserving agent of life, animality par excellence, the most beautiful, the most singular phenomenon of nature.
>
> In the living body, sensibility is the property of certain parts to perceive impressions of external objects, and to produce, as a consequence, movements proportional to the degree of intensity of that perception.[18]

Diderot, elsewhere in the *Encyclopédie*, defined sensibility simply as that which opposes death[19]; the term became synonymous with the 'vital principle'.[20]

In brief, the Montpellier vitalists' influence began in the late 1740s and 1750s. Determined to undermine the 'ordinary' medicine of their day, Bordeu, Venel, and Barthez (among others) moved to Paris. They went 'to school alongside Diderot, d'Holbach, and Rousseau at the Jardin Royal', and loosely joined forces with the *philosophes*, 'Bordeu in particular [making] a powerful impression on the Encyclopaedist circle'.[21] By the mid-eighteenth century, Montpellier vitalists were active in Parisian medical journalism and publishing, in the court, and in the salons, particularly d'Holbach's. Though they never 'sought to lead the *philosophes* in their campaigns against religious and philosophical tradition […] there can be no doubt that they left their mark on the Holbachian coterie'.[22] Their influence on the

---

[16] Packham 2012, 1.

[17] Vila 1998; Rey 2000; Williams 1994, 2003.

[18] '*la faculté de sentir, le principe sensitif, ou le sentiment même des parties, la base & l'agent conservateur de la vie, l'animalité par excellence, le plus beau, le plus singulier phénomène de la nature.*

La sensibilité est dans le corps vivant, une propriété qu'ont certaines parties de percevoir les impressions des objets externes, & de produire en conséquence des mouvemens proportionnés au degré d'intensité de cette perception'. Fouquet 1765, 38. My thanks to Kim Hajek for assistance with the translations.

[19] Diderot 1755, 782.

[20] Wolfe and Terada 2008, 540.

[21] Williams 2003, 147.

[22] Williams 2003, 131. More generally, see Williams 2003, 124–138, 147; Rey 2000, 2–3; Vila 1998, 45–51.

Encyclopédie is significant, particularity the contributions of Ménuret and Fouquet to the 1765 volumes.[23]

The importance of vitalism in the Scottish Enlightenment has been less recognised in contemporary scholarship, though we may note the recent study by Catherine Packham, who has drawn attention to the extent of vitalist medical thought in Edinburgh.[24] Two figures are of particular note here: Robert Whytt, notably for his 1751 *An Essay on the Vital and Other Involuntary Motions of Animals*, and George Cheyne for *The English Malady*.[25] The purpose of Packham's study was to link vitalism to emerging literary trends and the novel of sensibility; in doing so, she paralleled Anne Vila's 1998 study on the French Enlightenment. It remains the case, however, that the significance of vitalism for the history of Anglophone philosophy remains under-appreciated. And while the impact of vitalism on figures such as Adam Smith and David Hume has been noted, it has not yet received wide attention.[26]

Perhaps the most significant feature of vitalist medicine was its rejection of mechanist or corpuscularian theories of matter, which it was felt could not account for phenomena associated with living matter.[27] Broadly, the vitalists sought to bridge the mind-matter dichotomy, positing in living matter the existence of active, self-activating, or self-organising forces with their origin in the active powers of matter itself.[28] It is generally understood that there were two major sources for vitalist medical thought. First, the new physiological model of the body which emerged in the 1740s and 1750s in the experimental physiology of Albrecht von Haller, and his twin concepts of irritability and sensibility.[29] Haller sought to develop an empirically grounded understanding of organic structures and their functions. His chief concern was to demonstrate experimentally the existence of irritability, understood as the capacity of muscular fibres to contract upon stimulation. He distinguished this motile property from that of feeling, which he called sensibility and which he linked to the nervous fibres and associated with the soul. This distinction was not respected by the vitalist tradition, which merged the two and increasingly took sensibility to be a singular property with two aspects.[30] Though indebted to Haller, the *montpelliérains* did not inherit his experimentalism, but preferred instead observation and reflection.[31] The second major source of medical vitalism was the animism of Georg Ernst Stahl, who described the living

---

[23] Williams 2003, 123. For more on this see Lloyd, Chap. 9.
[24] Packham 2012.
[25] Packham 2012, 5–7. See also Packham 2012, 103–121; Wolfe, Chap. 8.
[26] See Cunningham 2007; Packham 2002.
[27] Reill 2005, 5, 33–70. See also Wolfe, Chap. 8.
[28] Reill 2005, 6–7. See also Gaukroger 2010, 387–420. For a detailed analysis of vitalist theories of matter, see Wolfe, Chap. 8.
[29] Boury 2008; Vila 1998, 13.
[30] See Wolfe, Chap. 8.
[31] Vila 1998, 46; Boury 2008, 530.

body in terms of an innate or internal force.[32] Boissier de Sauvages followed Stahl's lead in his lectures in Montpellier; Bordeu and Barthez were his students.[33] It is important to note, however, that the *montpelliérains* distanced themselves from the metaphysical aspects of Stahl's doctrine even as they agreed with his insistence on the singularity of life.[34] Rather than making strong metaphysical claims, they 'preferred ambiguous or disjunctive hypothetical statements when speaking about the relationship of emergent properties to those on which they supervene, as we would put it today'.[35] In this, there is an explicit affinity with Newtonian understandings of gravity[36]; the relevant biological/vital property is treated epistemologically; it is in this sense, that Charles Wolfe can speak of vitalism without metaphysics.[37] The tendency to avoid strong metaphysical claims regarding the precise nature of the vital force is part of what facilitated the wide spread of the discourse of sensibility over the eighteenth century.[38]

Vitalist medicine's understanding of the body of sensibility drew heavily upon, and interacted with, sensationist epistemologies. Although Locke did not make it a significant aspect of his *Essay Concerning Human Understanding*, concentrating instead on philosophical analysis of the problem in the manner of Descartes, the effect of his epistemology was to introduce the sensing or sensitive body to the problem of knowledge; Locke opened the door which allowed the problem of knowledge to become a question for philosophical medicine.[39] Locke's influence is multifaceted and contested, and 'it is difficult to overestimate the historical importance of Locke's theory of belief for the eighteenth century'.[40] Developing out of Lockean epistemology and taken up broadly in the Scottish Enlightenment, sensationist epistemology was widely influential in France, where it was adopted and systematised by Condillac, among others.[41] 'Sense' and 'sensation' were the key concepts here. Famously, Locke argued that the mind is initially blank and that all knowledge came in the first instance from the senses.[42] Sensations were associated (though not exclusively) with simple ideas: light and colour, which came through the eyes, noises which arrive only through the ears, and so on.[43] There is, however,

---

[32] French 1990. See also Reill 2005, 9–10, 61.
[33] Martin 1990; Cheung 2008, 495–496, 502.
[34] Vila 1998, 43; Cheung 2008.
[35] Kaitaro 2008, 583.
[36] Wolfe and Terada 2008, 542.
[37] Wolfe 2008.
[38] Packham 2012, 4.
[39] Vila 1998, 44; Suzuki 1995, 336–337. Suzuki argues that Locke became heavily influential on medical discourses in the late eighteenth century, though not earlier. See also Vermeir and Deckard 2012, 9, 12.
[40] Kuehn 2006, 391. See also Tipton 1996, 69–70.
[41] Brown 1996, 12; Kuehn 2006, 399; Knight 1968, 8–17.
[42] Tipton 1996, 74–75. See also Vermeir and Deckard 2012, 10.
[43] Locke 1690/1849, 63.

an important qualification to make here. Notwithstanding his attack on innate ideas or principles, Locke, in fact, held that there were two sources of knowledge, namely sensation and reflection, with reflection derivative from sensation. Locke defined reflection as 'internal sense', and 'what internal sensations [...] produce in us we may thence form to ourselves the ideas of our passions'.[44] This ambiguity between 'sensing' and 'thinking' was also present in Malebranche, for whom 'judgements and inferences, just like ideas themselves, are not *made* so much as *perceived*: they are themselves pure perceptions'.[45] The thinking body and the sensing body began to merge.

The increased importance that sensationist epistemology accorded 'sense' and 'sensation' gave impetus to the move to ground morality in a *moral* sense.[46] In broad terms, against those holding that morality was based either on self-interest (Hobbes, Mandeville, and later in France, Helvétius), or on reason (Cudworth), moral sense theorists held morality to be founded on a disinterested moral 'sense' or 'sentiment'.[47] As the tradition developed, so too did ideas of how the moral sense worked. The Earl of Shaftesbury, generally taken to mark the start of the tradition, was a moral realist: he understood the moral sense to pick out real characteristics in another person. This made easy work of the notion of a disinterested moral sense; moral judgements operated as any other sense perception and consequently allowed immediate and disinterested awareness of moral properties.[48] We can see here that the movement between the terms 'sense' or 'sensation', and 'sentiment', understood as feeling and moral judgement, was not accidental.[49] In Shaftesbury and the moral sense tradition following him, there was a strong relationship between moral and aesthetic judgements, as both were understood to be immediate and disinterested.[50] However, Shaftesbury's moral realism was weakened by subsequent moral sense theorists. Where, for Shaftesbury, the moral sense responded to a platonic notion of the harmonious and virtuous soul of the other, Hutcheson held that what the moral sense approves of in the other was benevolence.[51] In the culminating work of the

---

[44] Locke 1690/1849, 144. It is worth noting that the meaning of Locke's 'internal sense' easily blended into that of 'sentiment' understood in terms of the passions, for example in the context of the sentimental novel; there was a certain fluidity in the key concepts within the discourse of sensibility.

[45] Taylor, Chap. 4.

[46] As well as being a feature of scholarship on vitalism, the relationship between moral sense theory and the sentimental novel has been much discussed. See Mullan 1996, 249; Mullan 1988; Keymer 2005, 578–579; Brewer 2009, 22; Ellis 1996, 9–14. Though the scope of his book is much broader than the narrower themes discussed here, see too, Lamb 2009. See also Vermeir and Deckard 2012, 22.

[47] Norton and Kuehn 2006.

[48] Irwin 2008, 354, 362, 419; Norton and Kuehn 2006, 946.

[49] van Sant 1993, 7.

[50] Irwin 2008, 355; Radcliffe 2002, 456.

[51] Radcliffe 2002, 463. On the relationship between Shaftesbury and Locke, see Yaffe 2002, 425.

moral sense tradition, Adam Smith's 1759 *Theory of Moral Sentiments*,[52] 'sympathy' was the key concept.[53] That is, whether a trait was a virtue or a vice depended on whether one responded to it sympathetically, with fellow feeling, or with a reproduction of the feeling. Imagination was the central moral operator for Smith; it allowed one to place oneself in the other's situation and so feel for them sympathetically; thus Smith was able to explain the moral sense without invoking an independent dedicated faculty.[54] The link to the sentimental novel is clear, and literature takes a central place in Smith's *Theory*.[55]

A similar development of ideas occurred in France. Etienne Simon de Gamaches's 1708 *Le Système du cœur* outlined the operation of sympathy in much the same way as Smith later would, while Louis-Jean Levésque de Pouilly's 1747 *Théorie des sentimens agréables* had a significant influence on Smith. Other notable texts included Louis-Sébastien Mercier's 1767 novel *La Sympathie* and, of course, Rousseau's 1782 *Confessions*, in which sympathetic feeling was a highly prominent theme.[56] As for direct appropriation of the Anglophone moral sense tradition, Diderot was an early translator of Shaftesbury,[57] the *Encyclopédie* article 'Sens moral' quotes Hutcheson directly,[58] and on its publication, Smith's *Theory* received an immediate reaction in France.[59] Thus, *Encyclopédie* articles such as 'Sensibilité, (*Morale*)' and 'Sympathie, (*Physiolog.*)' are highly continuous with Smith, even if his name is not mentioned directly.[60] A final link in the chain is *Les Lettres sur la sympathie* by Sophie de Grouchy (1798), a new translation of Smith's text, accompanied by an extensive commentary. For de Grouchy, sympathy was not a product of the imagination, but instead something felt or sensed (*senti*); sympathy became a property of matter.[61] The link with vitalist theories of embodiment is clear.

We are now in a position to understand the significance of aesthetics for the Enlightenment. In this period in which the novel was stabilising as a genre, the sentimental novel was the dominant literary form.[62] Its dominant characteristic was the presentation of delicate or refined affective states or 'sentiments', particularly of tender feelings with regard to the plight of others. This 'language of feeling' marked

---

[52] Irwin 2008, 679.
[53] It was a feature, too, of Hume's moral theory. See Taylor, Chap. 4.
[54] Irwin 2008, 682–684.
[55] Fleischacker 2002, 509.
[56] Bernier 2010; Vervacke et al. 2007.
[57] Brewer 1993, 60–74.
[58] Jaucourt 1765a, 28.
[59] Bernier 2010, 1.
[60] Jaucourt, Louis de 1765b, c.
[61] Bernier 2010, 13–14.
[62] The sentimental novel has been the subject of much scholarly attention, including Ellis 1996; van Sant 1993; Mullan 1988; Vila 1998; Lamb 2009; Stewart 2010; Festa 2006; Barker-Benfield 1992. For a good summary of the development of the twentieth-century critical literature on the novel of sensibility, see Gaston 2010.

a concern with the interiority of the subject, both the subject as constructed within the text and the reader as positioned by the text.[63] For John Brewer,

> The poetics of sensibility depended upon the opening up of the private realm—interior feelings, emotional affect, intimate and familial friendship, the transactions of the home, the business of the closet, parlour, even bedroom—to public view. And it also privileged intimate and personal expression as true feeling untainted by a worldly desire for wealth and fame—hence the fiction of the editor employed by novelists like Richardson who posed as those who did not so much write as bring into the world a private, familiar correspondence.[64]

The novel of sensibility in its Anglophone incarnation was most famously realised in two authors. First, Samuel Richardson who, especially in his 1748 *Clarissa*, 'established "sentiment" as the very purpose of reading fiction'.[65] The sentimental novel was the more or less direct inheritor of the reformed domestic novel initiated by him.[66] Second, Laurence Sterne, particularly with his 1768 *A Sentimental Journey Through France and Italy*. For the French context, Philip Stewart traces a development from the language of passions to that of sentiment through Prévost, Marivaux, and Crébillon, before arriving at the 'triumph of moral sentiment',[67] in works by Diderot (particularly *La Religieuse*) and Rousseau (notably *La Nouvelle Héloïse*), and by the critics/satirists, Laclos and Sade.[68]

Of the concepts which are central to this volume, 'sentimental' and 'sentiment' are here dominant. 'Sentimental' was used in the older English sense of showing refined and elevated feelings. This is reflected, too, in the French, where *sentiment* 'expresses itself figuratively through the spiritual domain, in the various perspectives of the soul considering things'.[69] The term came to be associated with the passions: 'sentiment expresses itself, too, in the code of the passions, and signifies tender affection, love'.[70] The novel of sensibility did not just focus on the passions, but also took 'sentiments' to be moral precepts.[71] Clearly, the very term 'sensibility' is also significant, although here, it did not obviously carry the meaning attributed by vitalist medicine. Rather, 'sensibility' in this literary sense developed out of a notion of 'delicacy'; the association was with sensuous delight, superiority of class, fragility or weakness of constitution, tenderness of feeling, and fastidiousness.[72] Finally, literature was considered a means by which 'sympathy' and the moral sense were trained, such that writing and reading became performances of affect. 'Sentimental texts appealed to the benevolent instincts of a virtuous reader, who might be expected to suffer with those of whom he or she read'.[73] Literary representations were held to

---

[63] Brewer 2009.
[64] Brewer 2009, 35.
[65] Mullan 1996, 245.
[66] Ellis 1996, 44.
[67] Stewart 2010.
[68] Vila 1998, 111–181, 226–258.
[69] Stewart 2010, 5.
[70] Stewart 2010, 8.
[71] Mullan 1996, 246.
[72] van Sant 1993, 3.
[73] Mullan 1996, 238. See also Brewer 2009, 29.

have the same effect as real experiences. In this sense, the sentimental novel, 'constitute[d] a training-ground for the sympathies from which readers would emerge newly equipped to put them benignly into practice'.[74] The close proximity to notions around moral sense theory is evident.

In this historical and historiographical survey, I have sought to provide a broad and brief topography of the discourse of sensibility, highlighting its clusters of intellectual cultures around vitalist medicine, sensationist epistemology, moral sense theory, and the sentimental novel. The survey is necessarily incomplete, however, omitting, for example, discussions of the rise of philanthropic organisations or the significance of European colonialism.[75] I have similarly passed over the particular situation of women within the discourse and the problems posed by what was thought to be their particular, delicate, and refined (read: often pathological) sensibility. One example here would be Mary Wollstonecraft's critique of the construction of women's 'sensibility' as separated from the 'sensible'.[76] This is perhaps the major lacuna of this collection. Nonetheless, this topography will serve as a background to the substantive chapters collected in this volume and to the themes which traverse them.

1.3 The Chapters

The essays in this volume are a series of localised studies which together work to bring into focus the Enlightenment's knowing body. In approaching an historical object such as the body of sensibility, there are two major modes of questioning which can inform the historian's reconstructive and interpretive task. First, the question of emergence: where did it come from? Or perhaps: what were the conditions of its emergence? Second, what was at stake when considering this object? The chapters in this collection proceed in terms of both of these questions, though each to differing degrees. As a whole, the collection proceeds from those chapters which focus more heavily on the question of emergence to those whose concern is primarily with issues at play within the discourse.

1.3.1 The Emergence of the Discourse

This collection begins with the question of the emergence of the discourse of sensibility and with a focus on sensibility in its literary or aesthetic modes. Complexifying those aspects of established historiography which see the discourse of sensibility as emerging out of Cambridge Platonism and responses to Hobbesian moralities of

---

[74] Keymer 2005, 576. See also Vermeir and Deckard 2012, 39.
[75] van Sant 1993, 27; Brewer 2009, 23; Ellis 1996, 17; Festa 2006, 2.
[76] Vermeir and Deckard 2012, 27–28.

self-interest is Bridget Orr's essay on 'Richard Steele and the Rise of Sentiment's Empire' (Chap. 2). Orr provides an account of the emergence of sentimental literature and modes of sentimental writing as the narration of interiority by arguing for greater recognition of the importance of the early eighteenth-century English theatre in the rise of sensibility. The argument focuses on the figure of Steele, on his marginalised and contested ethnic identity, and on the context of his 1721 sentimental comedy *The Conscious Lovers*. As Orr shows, this was a period in which British national identity was highly contested: united by Protestantism and a commitment to trade and to Empire, divided by religious and national differences. In this context, the pamphlet attacks on Steele centred on his ethnicity. He was presented as a fortune-hunting Irishman and was criticised for his passage from colonial obscurity to the nation's seat of power. An emphasis on his financial problems, his drunkenness, his ingratitude to patrons, and his alchemical projects was framed in terms of negative characterisations of his nationality.

Orr's chapter introduces two of the major themes which will traverse this volume. First, the chapter adds to our understandings of how, as the discourse of sensibility emerged, it brought with it a development in modes of scripting the self, such that this emergence can be seen as a key moment in the making of the modern psychological self. As sentimental modes of narrating interiority evolved—particularly as sentimental fiction matured under the pen of the period's two most iconic epistolary novelists, Richardson and Rousseau—they came to script a sympathetic subject. In the chapter, Orr specifically examines Steele's self-creation in terms of the literary subject of sensibility. When attacked in the pamphlets of the day for his marginalised ethnicity, Steele enacted his defence by scripting himself into a miniature sentimental narrative. But it was particularly in the theatre that Steele 'work[ed] to invent a sympathy machine' which would operate to produce a common and proper feeling in his diverse and often-divided audience. Steele defended his 'Englishness', in part, by appealing to the English 'sincerity of heart and innate honesty'. After Steele, 'dramatists repeatedly turned to pathos and sentiment in attempting to generate religious toleration, cultural rapprochement, and national reconciliation'.[77]

Second, Orr introduces the theme of the particular and the universal. It was theatre's capacity to raise strong bonds of common feeling in its audience which is central to Orr's chapter. Because of his own hybrid status, Steele was ideally situated to use the 'sympathy machine' of the theatre to originate literary techniques which managed the 'difference and distance' of various marginal identities, by appealing to a unified or singular national culture of 'universal' (albeit particularly English) feeling.[78] As Orr writes, 'The aim of the dramatist was to unite the audience in a sympathetic response to suffering virtue, the sign of such pity being tears'.

> By weeping, the audience demonstrated their incorporation of and assent to the sentimental norms modelled on stage. No response could distinguish one weeping spectator from

---

[77] All citations from Orr, Chap. 2.

[78] The anxiety created by the radical possibilities of this 'sympathy machine' is the theme of Otto's essay (Chap. 10).

another except an indifference which would mark the viewer as uncivil [...] The value of the sentimental drama was equivalent, as its novel union of dramatic elements succeeded in erasing ethnic as well as sectarian status or party differences in a temporary community of proper feeling.[79]

Orr's conclusion is striking, however. Even as sentimental drama worked to erase distinctions within what was a contested national identity, it also worked to confirm in the spectators a sense of national superiority which they constructed in terms of their particular possession of enlightened virtues and humane feeling. In other words, *universal* humane feeling came to be understood as a *particular* characteristic.

The aesthetics of sensibility also frame Brandon Chua and Justin Clemens's chapter 'Rochester's Libertine Poetry as Philosophical Education' (Chap. 3), though here the problematic involves libertine judgement and poetic writing. The chapter enters in a somewhat unexpected manner into the question of the relationship between Hobbes and the emerging cultures of sensibility. It shows that if the discourse of sensibility, including theories of moral sense, emerged as a reaction *against* Hobbes's ethics of self-interest, it also emerged from a problematic or 'conceptual abscess' *within* Hobbesianism. Restoration libertinism was a culture founded on Hobbes's overthrowing of traditional moral philosophy and his valorisation of the pleasure-seeking individual as the source of the social and political contract. It celebrated the Hobbesian levelling of men to a condition of total equality in the state of nature. But as identified by Chua and Clemens, this levelling also posed a question: 'if Thomas Hobbes is right about materiality and sovereignty, then what are the consequences for individual action'? Through a close reading of John Wilmot, Earl of Rochester's *A Ramble in St James's Park* (c. 1672–1675), the chapter reveals an emerging need for a theory of substantive ethics which went beyond Hobbes's emphasis on rational calculation and egoism. It shows, in particular, that within this theory, there arose a demand for new accounts of the functionings of erotic and moral judgement.

Part of Chua and Clemens's argument rests on a shift in Rochester's mode of expression from that of the (Hobbesian) philosophical treatise to libertine poetry; in a similar vein, in my chapter (Chap. 9), I stress that within the discourse of sensibility, the philosophical novel was often the genre of choice. But sensibility and its turn to the body and to affect did not just emerge from a literary/aesthetic reaction to philosophy; this movement was also foreshadowed in formal philosophical writings. Jordan Taylor's chapter 'Emotional sensations and the moral imagination in Malebranche' (Chap. 4) makes this clear. This chapter explores Malebranche's theory of perception and the passions, and its influence on David Hume. Central to understanding this influence is Malebranche's account of the body, which Taylor outlines, concentrating on Malebranche's treatment of sensory perception, the imagination, and the passions. Notwithstanding Malebranche's own deontological ethics, Taylor demonstrates he had available to him within his system an alternative ethics, which could have been based on his account of the embodied aspects of the

---

[79] Orr, Chap. 2.

mind and the sensations experienced in perception. Taylor argues that Hume shared Malebranche's theories of embodied passions, of the imagination, and of sympathy and the mind's natural inclination towards compassion; despite notable incompatibilities in their ethical commitments, the two philosophers had more in common than is often acknowledged.

The contribution of Taylor's chapter to this collection is dual. On the one hand, it presents an image of mechanistic theories of the body (even if Taylor stresses that this is not a 'mere rigid mechanism'). Malebranche's physiology, including his account of the passions, was highly continuous with that of Descartes, being constructed in terms of animal spirits which moved in the brain according to prior pathways or etchings. Such a conception stands in stark contrast to vitalist theories of matter which became central to the discourse of sensibility and which are explicated in other chapters in this volume, particularly Charles Wolfe's (Chap. 8). In terms of matter theory, therefore, Taylor's chapter illustrates notions which would be overturned by the emerging discourse of sensibility. But significantly—and this is the major theme of the chapter and its key contribution to scholarship on the emergence of sensibility—Taylor shows the way in which, rather than constituting a strict rupture with the theories preceding it, the discourse of sensibility arose from within those theories. The chapter argues this point with reference to the continuities between Malebranche and Hume; however, there are also very significant continuities with later aspects of French theory, which I discuss in Chap. 9. As Taylor argues, 'although Malebranche's physiology is incomplete, it can be read with an air of flexibility, a certain neurobiological agnosticism'; my own chapter traces some of the forms taken by this agnosticism almost a hundred years later even as the period's physiology became increasingly complete. That is, even as matter theory developed and became increasingly sophisticated, it often did so within an ostensibly orthodox occasionalism which, as I argue, continued to provide the homogeneity of a faculty of rationality and perhaps also the reassurance of religious orthodoxy.

The chapters by Chua and Clemens and by Taylor work against the idea of a simple development of the discourse of sensibility as grounded in a rejection of mechanism and Hobbesianism. What both chapters reveal is the demand for, and the emergence of, substantive theories of embodiment from *within* mechanistic physiologies. They show that out of a mechanistic/materialist problematic emerge theories of feeling, of what it is to feel, and thus of what it is to feel *better*.

1.3.2 *Within the Discourse of Sensibility*

Alexander Cook's chapter 'Feeling Better: Moral Sense and Sensibility in Enlightenment Thought' (Chap. 5) constitutes a hinge or pivot for this collection. In opening, Cook continues this volume's focus on the emergence of the discourse of sensibility, tracing the rise during the seventh century of the status of the body and of philosophical anthropologies which postulated that humans are driven by their passions and not by reason. The chapter examines some of the dynamics which drove the growing preoccupation with sensibility and discusses how these dynamics explain the central-but-contested status of sensibility, understood as a concept bridging nature and culture, embodiment and morality.

Cook's chapter begins to trace exactly what was at stake within the discourse of sensibility. He introduces the third of this volume's major themes, the 'centrifugal tendency' within the discourse, and concomitantly, the contested nature and inherent ambiguities of the concept of sensibility.

> We should expect many histories of a concept which the persecuted French materialist philosopher Claude-Adrien Helvétius sought to make the basis of a moral science and which the conservative British Evangelical Hannah More declared to be 'virtue's precious seed'. Sensibility in the eighteenth century was one of those protean terms whose intellectual prestige and fundamental ambiguities invited myriad forms of appropriation.[80]

For Cook, the contested nature of sensibility was a feature of the fact that, as the discourse grew, sensibility had to do 'ever more work—at both the intellectual level and the social one'.[81] Cook's chapter moves towards a specific discussion of the dualistic notion of sensibility in the writing of Rousseau. Specifically, Cook argues that the relationship between 'corporeal sensibility' and 'imaginative sensibility' remained in dispute[82]: sometimes there was a tension between the two, other times an accord.

Cook's chapter develops the theme of the particular and the universal by examining the manner in which sensibility was construed as a variable and unevenly distributed quality, specifically noting the particular sensibility of women and of different social classes. In question was how the unevenness of the distribution of sensibility highlighted tensions within the discourse: the differential degrees of sensibility were managed by a tendency to understand it as a (particular) historical development of a (universal) natural property. It was essential for those who linked sensibility to the human potential for moral development that it be a universal feature of human nature. But sensibility was also treated as a rare and precious commodity which was not evenly distributed and which required cultivating. Here, Cook foregrounds the fourth of this volume's major themes, that of affective pedagogy, construed as efforts 'to cultivate the capacity for "proper" feeling against the dangers of its alternatives'.[83]

The theme of affective pedagogy, and the associated anxieties concerning the health of the learned, is also central to Yasmin Haskell's chapter 'Physician, Heal Thyself!: Emotions and the Health of the Learned in Samuel Auguste André David Tissot (1728–1797) and Gerard Nicolaas Heerkens (1726–1801)' (Chap. 6). From the mid-century on, intellectuals constituted a distinct patient group in the period's medical literature, thought to suffer from a nervous constitution, poor hygiene, and

---

[80] Cook, Chap. 5, quoting More 1782/1785, 282.

[81] Cook, Chap. 5.

[82] That sensibility had both an active/imaginative and a passive/corporeal nature is a key theme of my own chapter (Chap. 9).

[83] Cook, Chap. 5.

unhealthy work habits.[84] Emblematic of this anxiety often associated with the cultivation of sensibility was the figure of Samuel-Auguste Tissot. Famous for his pessimism regarding Enlightenment paths to progress, Tissot emphasised their complexities, contingencies, and the dire risks to health taken by those who travel them. But it is the engagement with Tissot's (and to a lesser extent with Rousseau's) pessimism undertaken by Gerard Nicolaas Heerkens, which is the focus of Haskell's chapter. Heerkens was a Dutch humanist physician and author of the 1749/1790 Latin poem *De valetudine literatorum* The Health of Men of Letters. As Haskell notes, both Heerkens and Tissot covered similar ground, including being wary of such things as the dangers to health of abusing tobacco and tea, of changes in the weather, and of late-night study vigils. In parallel, they both advised frugal diets and daily exercise, and both railed against parents who ruined the health of their children with unrealistic educational expectations. But where for Tissot there was 'almost always something *pathological* about learning', Heerkens had a predominantly cheerful approach to it. Heerkens can thus be associated with the cultures of medical hygiene which 'advocated the medical enhancement of sensibility and the cultivation of learning for self- and societal improvement', and which were also represented in the period by figures such as Vandermonde and Le Camus.[85] The theme of the sensibility of the learned was very much in play here; the idea that 'the more sensitive the mind is [...] the sharper it usually is' could be found in many texts of the period and accorded with the discourse's broader understanding of the knowledge-seeker. Haskell concludes, however, by arguing that Heerkens is best represented in terms of ongoing traditions of an older style of intellectual life: in contrast to the obsessive erudition of Rousseau and Tissot, for Heerkens, the life of the knowledge-seeker, which certainly had its psychological and physical dangers, also had very real emotional and intellectual compensations. Heerkens, with Diderot, 'will not concede that learning is in itself a disease nor even an unhappy life choice'.[86]

Anne Vila further pursues this concern with the embodiment of the learned, in her chapter entitled '*Penseurs profonds*: Sensibility and the Knowledge-Seeker in Eighteenth-Century France' (Chap. 7), which concentrates on the particular type of the absorbed or deep thinker. Vila uses this figure as a means of reconstructing the ways in which sensibility was held to affect the mind and body during acts of intense thought. Rather than approaching the question of the knowing body through matter theory (as Wolfe does, Chap. 8) or epistemology (as I do, Chap. 9), Vila approaches the question through a study of the way the figure of the absorbed or deep thinker was constructed in the writing of the period. The key figure here is Diderot; in question,—continuing a theme established by Orr—the manner in which he scripted the subject of sensibility, especially in the *Rêve de d'Alembert* and the *Éléments de physiologie*, but also in his personal letters to Sophie Volland.

---

[84] See also Vila, Chap. 7.
[85] Le Camus is a significant figure in my own chapter, Chap. 9.
[86] Citations from Haskell, Chap. 6.

1 The Discourse of Sensibility: The Knowing Body in the Enlightenment

As Vila shows, thinking was understood in whole-of-body terms. And Tissot was not the only one who warned the public about the risks of intellectual labour; rather, over-zealous or misdirected indulgence in learning was the cause of widespread concern. It was commonly held among the *médecins philosophes* that deep meditation caused sensibility to be channelled away from other centres of sensibility and towards the brain. The repercussions for the rest of the body were inevitably deleterious. This view manifested in two prominent ways: first, in the notion that the stresses created by learning made scholars more susceptible to illness; and second, in concern that pleasure-driven absorption in scholarly pursuits prevented scholars from participating in the larger social realm. Nonetheless, Vila finds in Diderot, as Haskell does in Heerkens, an approach to intellectual labours which was marked by optimism rather than anxiety. The chapter argues that, while he noted the dangers raised prominently by many in the period, Diderot was principally intrigued, rather than worried or disapproving, by the state of sensory oblivion that seemed to occur when a person is lost in thought. He found considerable philosophical appeal in the state of extreme mental absorption, and ultimately considered such absorption and mind-wandering to be productive. This was especially the case when the ideas responsible for triggering the state involved abstract thinking and when the state allowed the thinker to make connections between ideas that led to the discovery of truth and beauty. What fascinated Diderot was not the unhealthiness of the thinker lost in thought. It was rather the ability of the body to continue to function like a 'well-integrated animal economy whose internal parts have their own sort of awareness or attentiveness to their surroundings, along with a capacity for discernment that insures both the self-preservation and the preservation of the whole'.[87]

Charles Wolfe, in his chapter 'Sensibility as Vital Force or as Property of Matter in Mid-Eighteenth-Century Debates' (Chap. 8) approaches the question of the knowing body by foregrounding the matter theories which underpinned the discourse of sensibility. He gives a topography of the types of theories which Chua and Clemens (Chap. 3) argued were revealed as necessary by Rochester's lived experience of Hobbesian libertinism, and which replaced mechanistic matter theories, including those of Descartes, Malebranche, and Locke. Wolfe traces the diverse conceptions of matter and sensibility and, in parallel with Cook's chapter, shows that, within the discourse of sensibility, there was no singular concept of sensibility. The chapter takes as emblematic three understandings of sensibility. First, the enhanced mechanist view associated with Haller which built up from the basic mechanistic property of irritability to the higher-level property of sensibility associated with the soul.[88] Second, a vitalist view proper, in which sensibility was fundamental. Here, the organism was construed as a sum of parts understood as 'little lives' or animal economies. This was the territory of Montpellier vitalism narrowly understood. In this context, Wolfe describes sensibility as a 'booster'

---

[87] Vila, Chap. 7.
[88] With reference to Taylor's chapter (Chap. 4), it is worth noting that this enhanced mechanist view may be compatible with Malebranche's physiology.

property—a higher-order property like thought, memory, or desire, which belonged to higher-level organisms: sensibility allowed matter theory to support a rich account of the phenomena of consciousness. Third, a materialist view which sought to combine the mechanistic rigour and explanatory power of the Hallerian approach with the monistic and metaphysically explosive potential of the vitalist approach. Diderot is Wolfe's example here. In his thought, sensibility came to be understood as a universal property of matter itself.

My own chapter, 'Sensibilité, Embodied Epistemology, and the French Enlightenment' (Chap. 9), examines the implications of the body of sensibility on the period's theory of knowledge. I show the effect on sensationist epistemology of the move away from a mechanistic/corpuscularian matter theory (as held by Locke) and toward the variants of matter theories outlined by Wolfe. Particularly important here was the dual aspect of sensibility as it was conceived of following the unification of Haller's two forces: sensibility was construed as passive and active, and a reconstruction of the theory of knowledge as it existed in the period must take into consideration both aspects. My chapter continues several of the themes which are central to the volume as a whole. It develops the theme of affective pedagogy and of the hygienic response to the problem of knowledge with a discussion of Antoine Le Camus's 1769 *La Médicine de l'esprit*. I further engage with the anxiety generated by the centrifugal tendencies within the discourse, especially in relation to Diderot's critique of what he took to be Helvétius's epistemological and moral relativism. Finally, I address the question of particular regions or types of sensibility, taking up the theme of the particular and the universal in my discussion of the period's characterisation of the particular affects of genius. As I demonstrate, the artistic genius was constructed in terms identical to the rational or scientific genius, and there was no clear distinction in this period between someone who was an acute observer of physical phenomena and an observer of moral phenomena. My chapter reintroduces the question of aesthetics into the collection by showing the proper epistemological importance of aesthetics in the period: in concluding, it marks a turn towards the last paper in the collection and a return to the first two.

As in my chapter, Locke is a significant figure in the final paper of the collection, Peter Otto's 'Sensibility in Ruins: Imagined Realities, Perception Machines, and the Problem of Experience in Modernity' (Chap. 10). The impact of Locke's epistemology ought not to be considered as restricted to epistemology narrowly construed, or even to theories and practices of science more broadly; rather its influence was spread across the eighteenth century and the discourse of sensibility. Specifically, Otto's chapter reveals the extent to which Lockeanism penetrated literary cultures. As historians of philosophy are well aware, Locke's representational realism was vulnerable to scepticism (paradigmatically in Berkeley and Hume). It is in this vulnerability that Otto locates a 'nascent gothic and an embryonic romantic sensibility, which dissolved the unified world of traditional metaphysics into multiple realities'.[89] Through an examination of Matthew Lewis's 1796 *The Monk* and its

---

[89] Otto, Chap. 10.

critical reception in both Coleridge's review and Mathias's satiric poem, Otto illustrates ways in which the sometimes unsettling effects of sensationism could be experienced. Central to the chapter is the idea that fictions structured society and regulated the passions of its members and that literature had a civilising, and thus also potentially a corrupting, function. It was this latter possibility which was the source of the anxiety generated by Lewis's novel. Otto's essay concentrates on the illicit pleasures and the overt anxiety created by Lewis's use of literature's affective techniques for its own sake and in the *absence* of moral training. For Coleridge and Mathias, *The Monk* was an affect machine which drew readers into an unreal world, rousing 'unnatural' rather than 'natural affections', and creating a sense of false identity and false consciousness. At stake was precisely the same ability of the aesthetic to function as a 'sympathy machine' that Orr shows as emerging in the theatre (Chap. 2). What Otto highlights is not, however, the ability of sensibility to generate uniform emotions in an audience, but rather the anxieties which are generated by the process of manipulation itself.

Otto completes the fifth and final major theme which develops across the chapters of this volume, namely, the anxiety generated by the discourse and by sensibility's double-sided, often Janus-faced, nature. In the case of the critical reception of *The Monk*, this anxiety arose from the '*manufacture* and consequent widespread *proliferation*' of affective states, which threatened the capacity for rational judgement. (Such a threat is a theme in my own chap. 9) The equivocal nature of the discourse of sensibility is again on show in Otto's chap. 9, with both conservatives and radicals leaning on sensibility's, and literature's, power to affect. On the one hand, traditional fictions enhanced the stability of the government and the 'empire of good sense', but on the other hand, sensibility could lead to the underground and perhaps illicit delights of 'consumer culture'. In Otto's view, it was the contradictory narratives which arose within the discourse of sensibility which caused anxiety: sensibility became at once 'villain, victim, and hero/heroine'. The secret complicity between these incompatible roles is foregrounded by *The Monk* and is the source of that horror (and delight) which is evidenced by Coleridge's reaction to it; '[s]ensibility, one might say, is always haunted by its own ruin'.[90]

1.4 Concluding Remarks

Each of the chapters in this volume stands alone as an individual study that reconstructs particular aspects of the discourse of sensibility and/or the conditions of its emergence. The act of gathering them together foregrounds several significant themes: First, this collection has reconstructed various modes by which the sympathetic subject was construed or scripted in modes which included the theatre, poetry, literature, and medical and philosophical treaties. It furthermore draws out those

---

[90] Otto, Chap. 10.

techniques of affective pedagogy which were implied by the medicalisation of the knowing body, and highlights the manner in which the body of sensibility was constructed as simultaneously particular and universal. Finally, it illustrates the 'centrifugal forces' which were at play within the discourse, and shows the anxiety which often accompanied these forces.

Beyond a methodological commitment to historical accuracy in reconstructing the discourse of sensibility, there is much here to engage contemporary intellectual interest. It is widely held that since the mid-1990s there has been an 'affective turn' in the humanities and social sciences. Although the following list is far from exhaustive, we may note such areas as phenomenological and post-phenomenological theories of embodiment, aspects of psychological and psychoanalytic theory, post-Foucauldian critiques of normalising power, materialist theories of the human/machine/inorganic, and particularly the tradition beginning with Spinoza and leading into cultural studies/critical theory via the work of Gilles Deleuze and Brian Massumi.[91] Affect theory has rediscovered many of the themes which were alive in the period which forms the subject of this volume. Or perhaps we can say it has *reinvented* many of these themes: affect theory has drawn heavily from late-twentieth-century French theory, but intriguingly, this was a period in which thought was *not* significantly influenced by its eighteenth-century predecessors, either French or Anglophone, but rather was predominantly influenced by trends in the Continental 'rationalist' tradition and in German idealism.

This volume moves to reconstruct the Enlightenment discourse of sensibility and the manner in which it shaped the persona of the knowledge-seeker and created the body of sensibility. The discourse introduced a new model of the thinking process that, although not necessarily materialist in its philosophical orientation, was nonetheless grounded in the body in a dynamic way. The discourse of sensibility unified aspects of Enlightenment thought that have been treated as disparate, subsuming, for example, the period's significant metaphysical differences. As this volume shows, the key concepts involved in the construction of the body were ambiguous and often contested. As such, the differences—rather than the common features of a discourse within which the differences were situated—have typically been the focus of research into the period's thought. This is a tendency which has, for example, artificially separated the radical Enlightenment from its more conservative counterparts. The task of reconstruction is not completed here, but it is my hope that this volume makes a contribution to the project by drawing attention, in each of the chapters, to specific moments in the discourse of sensibility and by foregrounding, as a collection, the interdisciplinary, international, and inter-textual nature of the knowing body in the Enlightenment.

Acknowledgments For their assistance in preparing this chapter, I would like to thank Peter Cryle, Kim Hajek, Peter Otto, and the collected members of the Centre for the History of European Discourses at The University of Queensland who kindly read and critiqued a previous version of the chapter.

---

[91] Lloyd and La Caze 2011; Seigworth and Gregg 2010; Leys 2011.

References

Barker-Benfield, Graham J. 1992. *The culture of sensibility: Sex and society in eighteenth-century Britain*. Chicago: University of Chicago Press.
Bernier, Marc André. 2010. Présentation. Les métamorphoses de la sympathie au siècle des Lumières. In *Les Lettres sur la sympathie (1798) de Sophie de Grouchy. Philosophie morale et réforme sociale*, ed. Marc André Bernier and Deidre Dawson, 1–27. Oxford: Voltaire Foundation, SVEC Collection.
Boury, Dominique. 2008. Irritability and sensibility: Key concepts in assessing the medical doctrines of Haller and Bordeu. *Science in Context* 21(4): 521–535.
Brewer, Daniel. 1993. *The discourse of enlightenment in eighteenth-century France: Diderot and the art of philosophizing*. Cambridge: Cambridge University Press.
Brewer, John. 2009. Sentiment and sensibility. In *The Cambridge history of English romantic literature*, ed. James Chandler, 21–44. Cambridge: Cambridge University Press.
Brown, Stuart. 1996. Introduction. In *British philosophy and the age of enlightenment*, Routledge history of philosophy, vol. 5, ed. Brown Stuart, 1–19. London/New York: Routledge.
Cheung, Tobias. 2008. Regulating agents, functional interactions, and stimulus-reaction-schemes: The concept of 'Organism' in the organic system theories of Stahl, Bordeu, and Barthez. *Science in Context* 21(4): 495–519.
Cunningham, Andrew. 2007. Hume's vitalism and its implications. *British Journal for the History of Philosophy* 15(1): 59–73.
Diderot, Denis. 1755. Epicuréisme ou Epicurisme. In *Encyclopédie ou Dictionnaire raisonné des arts et des métiers*, 35 vols., vol. 5, ed. Denis Diderot and Jean Le Rond D'Alembert, 779–785. Paris: Briasson, David, Le Breton & Durand.
Ellis, Markman. 1996. *The politics of sensibility: Race, gender and commerce in the sentimental novel*. Cambridge: Cambridge University Press.
Festa, Lynn. 2006. *Sentimental figures of empire in eighteenth-century Britain and France*. Baltimore: The Johns Hopkins University Press.
Fleischacker, Samuel. 2002. Adam Smith. In *A companion to early modern philosophy*, ed. Steven Nadler, 505–526. Malden: Blackwell Publishing.
Foucault, Michel. 2004. *The Archaeology of Knowledge*. Trans. A.M. Sheridan Smith. London/New York: Routledge Classics.
Fouquet, Henri. 1765. Sensibilité, Sentiment (Médecine). In *Encyclopédie ou Dictionnaire raisonné des arts et des métiers*, 35 vols., vol. 15, ed. Denis Diderot and Jean Le Rond D'Alembert, 38–52. Paris: Briasson, David, Le Breton & Durand.
Frazer, Michael L. 2010. *The enlightenment of sympathy*. Oxford: Oxford University Press.
French, Roger. 1990. Sickness and the soul: Stahl, Hoffmann and Sauvages on pathology. In *The medical enlightenment of the eighteenth century*, ed. Andrew Cunningham and Roger French, 88–110. Cambridge: Cambridge University Press.
Gaston, Sean. 2010. The impossibility of sympathy. *The Eighteenth Century* 51(1–2): 129–152.
Gaukroger, Stephen. 2010. *The collapse of mechanism and the rise of sensibility: Science and the shaping of modernity, 1680–1760*. Oxford: Oxford University Press.
Irwin, Terence. 2008. *From Suarez to Rousseau*, The development of ethics: A historical and critical study, vol. 2. Oxford: Oxford University Press.
Jaucourt, Louis de. 1765a. Sens moral. In *Encyclopédie ou Dictionnaire raisonné des arts et des métiers*, 35 vols., vol. 15, ed. Denis Diderot and Jean Le Rond D'Alembert, 28–29. Paris: Briasson, David, Le Breton & Durand.
Jaucourt, Louis de. 1765b. Sensibilité (Morale). In *Encyclopédie ou Dictionnaire raisonné des arts et des métiers*, 35 vols., vol. 15, ed. Denis Diderot and Jean Le Rond D'Alembert, 52. Paris: Briasson, David, Le Breton & Durand.
Jaucourt, Louis de. 1765c. Sympathie, (Physiolog.). In *Encyclopédie ou Dictionnaire raisonné des arts et des métiers*, 35 vols., vol. 15, ed. Denis Diderot and Jean Le Rond D'Alembert, 736–740. Paris: Briasson, David, Le Breton & Durand.

Kaitaro, Timo. 2008. Can matter mark the hours? Eighteenth-century vitalist materialism and functional properties. *Science in Context* 21(4): 581–592.

Keymer, Thomas. 2005. Sentimental fiction: Ethics, social critique and philosophy. In *The Cambridge history of English literature, 1660–1780*, ed. John Richetti, 572–601. Cambridge: Cambridge University Press.

Knight, Isabel. 1968. *The geometric spirit: The Abbé de Condillac and the French enlightenment*. New Haven/London: Yale University Press.

Kuehn, Manfred. 2006. Knowledge and belief. In *The Cambridge history of eighteenth-century philosophy*, 2 vols., vol. 1, ed. Knud Haakonssen, 389–425. Cambridge: Cambridge University Press.

Lamb, Jonathan. 2009. *The evolution of sympathy in the long eighteenth century*. London: Pickering & Chatto.

Leys, Ruth. 2011. The turn to affect: A critique. *Critical Inquiry* 37(3): 434–472.

Lloyd, Henry Martyn, and Marguerite La Caze. 2011. Editors' introduction: Philosophy and the 'affective turn'. *Parrhesia: A Journal of Critical Philosophy* 13: 1–13.

Locke, John. 1690/1849. *An essay concerning human understanding*, 30th ed. London: William Tegg & Co.

Martin, Julian. 1990. Sauvages's nosology: Medical enlightenment in Montpellier. In *The medical enlightenment of the eighteenth century*, ed. Andrew Cunningham and Roger French, 111–137. Cambridge: Cambridge University Press.

More, Hannah. 1782/1785. *Sacred dramas: Chiefly intended for young persons ... to which is added, sensibility, a poem*. Dublin: P. Byrne.

Mullan, John. 1988. *Sentiment and sociability: The language of feeling in the eighteenth century*. Oxford: Clarendon Press.

Mullan, John. 1996. Sentimental novels. In *The Cambridge companion to the eighteenth-century novel*, ed. John Richetti, 236–254. Cambridge: Cambridge University Press.

Norton, David Fate, and Manfred Kuehn. 2006. The foundations of morality. In *The Cambridge history of eighteenth-century philosophy*, 2 vols., vol. 2, ed. Knud Haakonssen, 941–986. Cambridge: Cambridge University Press.

Packham, Catherine. 2002. The physiology of political economy: Vitalism and Adam Smith's wealth of nations. *Journal of the History of Ideas* 63(4): 465–481.

Packham, Catherine. 2012. *Eighteenth-century vitalism: Bodies, culture, politics*. Basingstoke: Palgrave Macmillan.

Radcliffe, Elizabeth S. 2002. Francis Hutcheson. In *A companion to early modern philosophy*, ed. Steven Nadler, 456–468. Malden: Blackwell Publishing.

Reill, Peter Hanns. 2005. *Vitalizing nature in the enlightenment*. Berkeley: University of California Press.

Rey, Roselyne. 2000. *Naissance et développement du vitalisme en France de la deuxième moitié du 18e siècle à la fin du premier empire*. Oxford: Voltaire Foundation.

Riskin, Jessica. 2002. *Science in the age of sensibility: The sentimental empiricists of the French enlightenment*. Chicago: University of Chicago Press.

Seigworth, Gregory J., and Melissa Gregg. 2010. An inventory of shimmers. In *The affect theory reader*, ed. Gregory J. Seigworth and Melissa Gregg, 1–25. Durham/London: Duke University Press.

Stewart, Philip. 2010. *L'Invention du sentiment: Roman et economie affective au XVIIIe siècle*. Oxford: Voltaire Foundation, SVEC Collection.

Suzuki, Akihito. 1995. Anti-Lockean enlightenment? Mind and body in early eighteenth-century English medicine. In *Medicine in the enlightenment*, ed. Roy Porter, 336–359. Amsterdam: Rodopi.

Tipton, Ian. 1996. Locke: Knowledge and its limits. In *British Philosophy and the age of enlightenment*, Routledge history of philosophy, vol. 5, ed. Brown Stuart, 69–95. London/New York: Routledge.

van Sant, Ann Jessie. 1993. *Eighteenth-century sensibility and the novel*. Cambridge: Cambridge University Press.

Vermeir, Koen, and Michael Funk Deckard. 2012. Philosophical enquiries into the science of sensibility: An introductory essay. In *The science of sensibility: Reading Burke's philosophical enquiry*, ed. Koen Vermeir and Michael Funk Deckard, 3–56. Dordrecht/New York: Springer.

Vervacke, Sabrina, Thierry Belleguic, and Éric Van der Schueren. 2007. *Les Discours de la sympathie: Enquête sur une notion de l'âge classique à la modernité*. Lévis: PUL.

Vila, Anne C. 1998. *Enlightenment and pathology. Sensibility in the literature and medicine of eighteenth-century France*. Baltimore: Johns Hopkins University Press.

Williams, Elizabeth A. 1994. *The physical and the moral: Anthropology, physiology and philosophical medicine in France, 1750–1850*. Cambridge: Cambridge University Press.

Williams, Elizabeth A. 2003. *A cultural history of medical vitalism in enlightenment Montpellier*. Aldershot: Ashgate.

Wolfe, Charles T. 2008. Introduction: Vitalism without metaphysics? Medical vitalism in the enlightenment. *Science in Context* 21(4): 461–463.

Wolfe, Charles T., and Motoichi Terada. 2008. The animal economy as object and program in Montpellier vitalism. *Science in Context* 21(4): 537–579.

Yaffe, Gideon. 2002. Earl of Shaftesbury. In *A companion to early modern philosophy*, ed. Steven Nadler, 425–436. Malden: Blackwell Publishing.

Chapter 2
Richard Steele and the Rise of Sentiment's Empire

Bridget Orr

Abstract Literary historians and critics of sentiment have recently begun exploring the way in which sentimental modes in fiction and poetry were imbricated in the development of colonial and imperial subjectivities in eighteenth-century Britain. Such accounts either identify the discursive origins of sentimental literature in Cambridge platonism and moral sense philosophy, or simply bracket out the question of beginnings. By contrast, this chapter will stress the emergence of sentiment in the theatre, in late Stuart pathetic drama, reworked as sentimental comedy and domestic tragedy by Whig ideologues such as Richard Steele. This revision enables us to see the ways in which national and imperial sympathies were modelled and mobilised in dramatic contexts prior to their later invocation in narrative and poetic texts. Notably, the chapter argues that it was by watching theatrical performances of the emergent modes of pathetic and sentimental drama that audiences learned to become unified by common, national feeling.

Recent work on Sterne and other sentimental novelists has been much concerned with the imbrication of sentimental fiction and empire. The connection between abolitionism and sentiment has long been familiar, canvassed many years ago by Wylie Sypher and more recently by Markman Ellis, Deirdre Coleman, and others.[1] But the literary historical claims about sentiment's cultural functions have become much more extensive, most strikingly perhaps in Lynn Festa's *Sentimental Figures of Empire*, in which she argues that sentimentality replaced epic in the eighteenth century as the dominant literary mode of empire by magnifying and mystifying colonial relations and by

---

[1] Sypher 1969 is still an important survey of the relevant literature. See also the more recent studies, Ellis 1996; Coleman 2005; Carey 2005. For a recent discussion of the role of abolitionism in relation to national character, see Swaminathan 2009.

B. Orr (✉)
Department of English, Vanderbilt University, Benson Hall 331, Nashville 37235, TN, USA
e-mail: bridget.e.orr@vanderbilt.edu

generating the tropes that render relations with distant others thinkable.[2] For Festa, sentimentality structured flows of affect between metropolitan subjects and colonial objects, not only by refashioning scenes of violence, but by facilitating the constitution of the modern (metropolitan) self through repeated acts of emotional piracy.

Leaving aside Festa's implicit dismissal of the role played in the constitution of national and imperial culture by such genres as history plays or dramas of state and poetry, her account (like those of other recent scholars) begs the question of sentiment's origins as a literary mode in the theatre.[3] While she does note that sentimental tropes, characters, and plots migrate across generic boundaries as well as national borders, her examples of such migration are *Oroonoko, Inkle and Yarico,* and *Paul et Virginie,* all of which 'had a popular after-life on the stage', underscoring her claim that it was the novel that 'was in the vanguard of sentimental fashion'.[4]

There is, however, an alternative way to begin the story of sentiment and empire. To do so requires returning to an older scholarship in which the origins of the sentimental discourse are shown to lie as much in the theatre as in the formulations of the Cambridge Platonists or Lord Shaftesbury.[5] The theatrical experiments made by Richard Steele reveal an 'Englishman born in Dublin' creating dramatic vehicles intended to meld together the heterogeneous audiences who literally embodied the recently united kingdom. As Joseph Roach has suggested in *Cities of the Dead,* the late Stuart theatre was an early scene of imagined community, one in which one's fellow subjects are physically proximate rather than virtually united by the simultaneous, but spatially distinct, consumption of print.[6] The early eighteenth-century theatre was understood by contemporaries to model the kingdom as a whole, its fractious heterogeneity as much as its aesthetic peculiarities an index of the nation's historical vicissitudes, but also its great particularity—political liberty and religious toleration. For Colley Cibber, Steele's ally in the promulgation of Whig dramaturgy, theatre's unique capacity to raise strong common feeling among divided spectators was the key to its national importance, as we see in his remarks on Addison's *Cato* (1713):

> When the Tragedy of *Cato* was first acted, let us call to mind the noble Spirit of Patriotism, which that Play then infus'd into the Breasts of a free People, that crowded to it; with what affecting Force, was that most elevated of Human Virtues recommended? Even the false Pretenders to it felt an unwilling Conviction, and made it a Point of Honour to be foremost, in their Approbation; and this too, at a time when the fermented Nation had their different Views of Government. Yet the sublime Sentiments of Liberty, in that venerable Character, rais'd, in every sensible Hearer such conscious Admiration, such compell'd Assent to the Conduct of a suffering Virtue, as even *demanded* two almost irreconcilable Parties to embrace, and join in their Applause of it. Now, not to take from the Merit of the Writer, had that Play never come to the Stage, how much of this valuable effect of it must have been lost?[7]

---

[2] Festa 2006, 2–8.

[3] For discussions of the role of other genres see Orr 2001; Kaul 2000; O'Quinn 2005.

[4] Festa 2006, 9. This argument assumes that the novel, like lyric poetry, has a privileged capacity to express and sustain interior affect.

[5] See Bernbaum 1958.

[6] Roach 1996.

[7] Cibber 1740/1968, 196. Characteristically, Cibber stresses the moving effect of the play in performance.

David Marshall has stressed the importance of theatrical modelling in Adam Smith's account of sentiment, a theoretical framework for social bonding which cultural historians now argue constitutes a characteristically British way of understanding national identity.[8] My contention is that it was by watching theatrical performances of the emergent modes of pathetic and sentimental drama that audiences learned to become unified by common, national feeling. As Steele pursued a career as a cultural impresario in late Stuart/early Hanoverian Britain, his Irish origins shaped his embrace of reforming Whiggery's projects of politeness and piety, a cultural and political programme which could include and manage the relations of individuals and groups divided by ethnicity, rank, religious belief, and political loyalties. In working to invent a sympathy machine which would reduce all his audience, generals included, to tears, Steele was not simply aiming at a reconciliation of trade and land, or the promulgation of middle-class morality. The union he sought to create was one which might sink ethnic or national differences in proper feeling, ideally in what Cibber called 'the noble Spirit of Patriotism'. To succeed in this venture, he turned to old plays and, in particular, the she-tragedies of Banks. In *The Conscious Lovers* (1722), he created a template which would be used repeatedly by dramatists seeking to recuperate all those heterogeneous groups who were part of, but marginal to, the United Kingdom and empire—Jacobites, Jews, the Irish, Scots, nabobs, creoles, and African slaves. The extent to which dramatic sentiment actually succeeded in that later, more extensive project of 'humanisation' is as much contested now as it was in the eighteenth century: George Boulukos argues that sentimental depictions of 'the grateful slave' actually made racist discourse conventional, while David Worrall argues that anti-slavery plays popularised the abolitionist juggernaut.[9] What is certain, however, is that eighteenth-century British dramatists repeatedly turned to pathos and sentiment in attempting to generate religious toleration, cultural rapprochement, and national reconciliation.

2.1 Anglo-Hibernus

Steele's investment in an expansive and inclusive United Kingdom was shaped by his own colourful colonial background. His grandfather's career as a dashing East India merchant commissioned by the East India Company (EIC) to open up the Persia trade, and celebrated as a friend of the Great Moghul, was memorialised by Coryat. Steele's grandmother and her children (one of whom was born in India)

---

[8] See Marshall 1986. Recent work on the way in which Scottish Enlightenment theories of sympathy created 'national feeling' in the later eighteenth century include Gottlieb 2007; Shields 2010.

[9] See Boulukos 2009; Worrall 2007. Boulukos argues that while the (trope of) the slave's gratitude humanised him/her, it suggested a willingness to accept subservience which implied inferiority, enabling ameliorationist arguments to become dominant at the expense of abolition. The argument is compelling, but fails to take account of the effect of reaction to the French Revolution. In *Harlequin Empire*, David Worrall demonstrates the huge reach of abolitionist drama and the greater acceptance of black actors on British stages in the late Georgian period.

were settled by the family patriarch on a plantation near Ballyinaskill and experienced a terrible siege in the castle there during the Irish Confederation uprising of 1641. Orphaned at ten, Steele was removed to school in England, but he never hid his antecedents, taking on the function of Steward, for example, during the memorial processions for the '41 rebellion.[10] His always shaky finances received a very substantial boost when he inherited a Barbadian plantation from his first wife—who had herself inherited when her brother was captured by French privateers. The estate, worth at least 850 lb a year, was worked by 200 slaves, and Steele sold it after his first wife's death to facilitate his second marriage.[11] He was friendly with several West Indians, and was the first subscriber to offer books to the library at the newly established Yale College in Connecticut.

Steele was thus connected to at least three main arenas of colonial activity, but by no means consistently in positions of profit or mastery. His Irish origins were an obvious source of vulnerability as he sought to establish himself as a figure of political, as well as cultural, authority. Along with his collaborator Addison, Steele is recognised as a crucial protagonist in the creation and extension of the public sphere through the enormously popular periodicals *The Tatler* and *The Spectator,* which the pair edited and largely composed. Fashioning rhetorical vehicles such as Mr. Spectator, a character who watches, mediates between, and sometimes criticises rival groups and manners, was essential to this project—by creating self-effacing observers who claimed to be impartial and disinterested, they made a reliable cultural arbiter. Unlike Addison, however, who achieved high office through patronage, Steele sought to become an increasingly visible political player. As part of this process, he gradually abandoned the increasingly transparent personae of Isaac Bickerstaffe, Tatler and Censor, Mr Spectator, and Nestor Ironside to appear in his later periodicals without disguise, under his own name as 'an Englishman born in Dublin'. No longer regarded as an impartial observer above the fray, but an engaged participant in political and cultural conflict, Steele became the subject of ever more vituperative attack, with the publication of *The Importance of Dunkirk Considere'd* (1713) in particular producing highly personal responses. Steele's financial problems, his drunkenness, his supposed ingratitude to patrons, and his alchemical projects were all canvassed, but the primary focus of these negative characterisations is his nationality.[12]

In several of the hostile accounts of Steele, the dramatist figures as a fortune-hunting stage Irishman, an Irishman, moreover, whose profitable marriage to a Barbadian heiress taints him with a certain creole arrogance—Defoe accuses him of addressing the Queen 'just as an imperious Planter at Barbadoes speaks to a Negro Slave.'[13] John Lacy attacks him in similar terms in *The Ecclesiastical and Political history of Whigland* (1714):

---

[10] See Knight 2009, 10.

[11] Details of the West Indies estate are in Aitken 1889, Vol. 1, 132–133.

[12] The most recent survey of this aspect of Steele's career is found in Knight 2009.

[13] Defoe 1713, 8.

> Many Years ago, *Don Ricardo* had Ingenuity enough to make his own Fortune, by that Qualification, which seems to be more particularly innate to him, than any of his Countrymen, who are famous for a constant and diligent Impudence, the Practice of which, in the most flagrant Degree, gives them a Dominion over the weak Sex; who are unable, tho' they even hate, to resist such violent and unnatural attacks as they usually make upon 'em, till they are forced to be their Wives, as the only Way to get Rid of 'em. A West-Indian Beauty, attacked in this Manner, gave herself and with her Person, more Mines of Gold, than would have made a plentiful Fortune for a worthier Mortal than *Don Ricardo*.[14]

Lacy goes on to claim that Steele's infidelity caused his wife's death, an accusation already canvassed by Delariviere Manley in her *New Atalantis*: 'thick set, his Eyes lost in his head, hanging Eye-Brows, broad Face and tallow Complexion [...] has an inexhaustible fund of Dissimulation, and does not bely the Country he was born in, which is fam'd for falsehood and Insincerity'.[15] In a dialogue with a Mrs. Tofts later in the *Memoirs*, the lady remarks to Don Phoebo (Steele) of his dead wife that 'Your Fame is not quite so clear in Reference to that ugly and odd Misfortune, that was so fatal to her, occasion'd by your Sister'.[16]

When focusing on Steele's activities as a political author, the pretentions of an Irishman to English identity and authority are a recurring trope—as William Wagstaffe writes in *A Letter from the facetious Doctor Andrew Tripe at Bath*: 'Sir, I more particularly remember they said of you [...] that you attempted to make an Englishman of Teague'.[17] Swift's is perhaps the most extreme instance of such exclusionary language, when at the end of *The Publick Spirit of the Whigs* he suggests that

> I agree with this Writer, that it is an idle thing in his Antagonists to trouble themselves upon the Articles of his Birth, Education or Fortune; for Whoever writes to his Sovereign, to whom he owes so many personal Obligations, I shall never enquire whether he be a GENTLEMAN BORN, but whether he be a HUMAN CREATURE.[18]

Occasionally, Steele fought back in kind. In his *Apology for Himself and his Writings* (1714) in which he defended himself from the charges of seditious libel which had led to his expulsion from the House of Commons early in 1714, he described Thomas Foley, an in-law of Harley's, who led the attack against him as follows:

> The Man I mean was of an Enormous Stature and Bulk, and had the Appearance, if I may speak so, of a Dwarf-Giant. His Complection Tawny, his Mein disturb'd, and the whole Man something particularly unfamiliar, disingenuous, and shocking to an English Constitution. I fancied, by his exotick Make and Colour, he might be descended from a Moor, and was some Purchase of our African, or other trading Company, which was manumised. This Man, thought I, was certainly bred in Servitude, and being now out of it, exerts all that he knows of Greatness in Insolence and Haughtiness.[19]

---

[14] Lacy 1714, 12.
[15] Manley 1720, Vol. 4, 302.
[16] Manley 1720, Vol. 4, 307.
[17] Wagstaffe 1714, 28.
[18] Swift 1714, 39.
[19] Steele 1714/1944, 295.

This invention of a fantastic history for Foley involving North African ancestry, enslavement, manumission, and illegitimate entry into Parliament reworks many of the tropes employed in the attacks on Steele himself: blackness, creole arrogance, profoundly un-English origins, and a passage from colonial obscurity to the nation's seat of power. The biography which Steele creates for his accuser mirrors the delegitimating narrative projected onto him by the pamphleteers. But while its primary target is personal, the passage also implies that the slave trade has the capacity fundamentally to disfigure and denature the English constitution, evoked here in both its bodily and political dimensions. Unlike the mutually enriching operations of *doux commerce*, maritime trade has here injected an alien body into the nation's political heart—a heart unnaturally hardened, Steele goes on to suggest, by his accuser's unjust cruelty. As the tide seemed to be turning in Steele's favour, 'The untam'd Creature stood up to turn off the merciful Inclination which he saw grow towards the Member accus'd', suppressing members' natural inclination to tenderness.[20] In the miniature sentimental narrative Steele has constructed for his readers, the manumitted slave's mimicry of his former masters' tyranny triumphs over the natural benevolence of the nation's representatives.

Steele continued his campaign against the Harleys in *The Lover*, in terms which seem to underscore his sensitivity to the specifically ethnic nature of the attacks in *The Examiner* and elsewhere. At the close of *Lover* number 14, March 27, 1714, for example, 'Ephraim Cattlesoap' concludes his account of 'the Exotick and Comick Designs of this unaccountable Race', the Crabtrees (Oxford, his brother Edward, and Foley), who are, he writes, '(according to their own different Accounts of their Parts and Births) occasionally Syrians, Egyptians, Saxons, Arabians, and everything but Welch, British, Scotch, or anything that is for the Interest of these Dominions'.[21] The obscurely and exotically descended Harleys are contrasted to those whose positively valenced British identities actually exclude the English. The valorisation and claim to a Cambro-British heritage starts to replace Steele's self-described English identity. In 1720, for example, he claims 'I was begot in Dublin by a Welsh gentleman upon a Scots Lady of Quality', reinscribing his suspiciously hybrid Anglo-Irishness as a rich compendium of the United Kingdom's ancient nations.[22]

Although biographers find the attacks distasteful and critics by and large disregard the topic, it would have been near impossible for Steele's ethnicity to have been ignored in the pamphlet wars of the 1710s. A brief inspection of the recent historiography of British national identity in the eighteenth century makes it clear why Steele's position was so confused and vulnerable. In *Britons*, Linda Colley argues that the eighteenth-century British were united by their Protestantism, their hostility to Catholic enemies, their commitment to trade, and their common interest

---

[20] Steele 1714/1944, 295.

[21] Steele 1959, 53–54.

[22] Steele apparently made the claim in a debate on classifying Irish cloth as a foreign manufacture. See Knight 2009, 10.

in imperial expansion.[23] As other historians have noted, however, she pays rather less attention to the religious and national or ethnic differences that continued to divide the Scots, Irish, Welsh, and English under the later Stuarts and Hanoverians. While nativist traditions drawing on Celtic pasts were variously deployed from the seventeenth century on to stress cultural distinctiveness, Colin Kidd has shown the extent to which various theological and antiquarian arguments about pan-European Gothicism provided rhetorical resources for those who sought to underline the essential historical unity of the British as well as their ancestral links to other modern Europeans.[24] As Gothicism was incorporated into popular Whig apologetics in the post-Revolutionary period, it provided a specifically historical justification for understanding the component kingdoms of the British Isles as the common inheritors of the Teutonic legacy of liberty. Steele himself provides a classic articulation of this *idée reçue* in *The Englishman* No. 28, from 8 December 1713:

> If LIBERTY be then so valuable, those Nations whose Government has appear'd to be founded upon Maxims the most conducive and necessary to its Preservation, though not conversant in the politer parts of Learning, are so far from being deserving to be stiled *Barbarous*, that they justly merit as glorious Panegyricks as ever came from the Mouth of *Tully* or *Demosthenes*.
>
> AMONGST those may be reckoned the ancient Inhabitants of the Northern Parts of *Europe*, out of which in different Ages have gushed those mighty swarms of *Goths*, *Vandals*, *Saxons, Angles, Franks, Huns, Danes* and *Normans*, which subdu'd all the Western Parts of *Europe*.
>
> THE grand Northern HIVE from whence they came, has by some Authors been stiled *Officina Gentium*, the Shop of Nations; and might with as much Justice have been called *Officina Libertatis*, the Shop of Liberty.[25]

In the case of those we now call the Anglo-Irish, including Steele and Swift, the sense of an English identity that survived transplantation to Ireland was even more distinct, as new settlers distinguished themselves from the 'mere Irish'—the dispossessed Catholics—despite the fact that the metropolitan English refused to recognise such differences.[26] The irony for the Anglo-Irish—like other settlers—was that in colonising, they became colonials.

Steele's investment in the Whig project thus has a cultural and political specificity conditioned by his Irish antecedents. His admiration for William III was arguably informed by a positive attachment to the idea that what mattered in a 'Christian hero' was that he was a hero of the Protestant interest and that his non-English origins were decidedly irrelevant to his role as national saviour. Writing in *The Englishman* No. 3, of 10 October 10 1713, Steele explains:

> When I say an *Englishman*, I mean every true Subject of her Majesty's Realms, the *Briton* of the North as well as he of the South; and know no Reason for saying *Englishman* instead of *Scotsman*, but that latter Appellation is drawn into the former from the

---

[23] Colley 1994.
[24] Kidd 1999.
[25] Steele 1955, 113.
[26] For a detailed discussion of this issue see Kidd 1999, Chap. 7–10.

Residence of the Queen in the Southern Part of *Great Britain*. I abhor the Distinction, and think it absolutely necessary for our mutual Honour and Safety, as far as it is possible, to abolish it. It is below the Sincerity of Heart and innate Honesty of a true *Englishman* to enter into a partial Friendship; and it is a Matter of Lamentation, to observe the cool Distance, that is maintained towards Men who have resigned great Immunities, and placed themselves irrevocably under the same Soveraignty with us, in order to our mutual Wealth, Glory and Happiness.[27]

But as we have seen, the 'mere English' did not necessarily share Steele's views, not in politics, nor in the playhouse. John Dennis believed the stage's contemporary decline was caused by venal and low-bred actor-managers supplanting theatrical management by nobles; however, he also cited demographic shifts in the audience consequent on the more general political and social changes which followed the golden age of the Restoration:

The Audiences were *English* all or most of them, audiences that understood what They saw and Heard; and we had the none of those shoals of exoticks, that came in by the Revolution, the union, and the *Hanover* Succession, which tho They were events that were necessary all, and without which we had been undone; yet they have hitherto had but an evil Influence upon the genuine Entertainments of the stage, and the studies and arts of Humanity.[28]

For in addition to the 'shoals of exoticks' who require a dumbed-down theatre, Dennis is incensed by the undiscriminating ignorance of successful military men and stock-jobbers:

a new and numerous Gentry has risen among us by the Return of our fleets from sea, of our Armies from the Continent, and from the wreck of the South Sea. All these will have their Diversions and their easie Partiality leads them against their own palpable interest to the Hundreds of Drury.[29]

For Dennis, a vicious cycle had emerged, whereby a newly heterogeneous and uneducated spectatorship was pandered to by equally low, ill-informed, and in Steele's case, non-English theatrical managers. In his 'Picture of Sir John Edgar', he describes Steele as follows:

He was a Gentleman born, Witness himself; of a very Honorable Family, certainly of a very Ancient one. For his Ancestors flourish'd in *Tipperary* long before the *English* ever set foot in *Ireland*. He has Testimony of this more Authentick than the *Herald's* office or than mere Human Testimony; for God has mark'd him more abundantly than *Cain*, and stamp'd his Native Country upon his Face, his Understanding, his Writings, his Actions, his Passions, and above all, his Vanity. The *Hibernian* Brogue is still upon all these, tho long Habitude and Length of Days has worn it from off his Tongue.[30]

It is these Tipperary origins which Dennis cites obsessively as the source of Steele's imputed shameless avarice, his philandering, his nonsensical projects, and his plagiary.

---

[27] Steele 1955, 14.
[28] Dennis 1725/1943, 276.
[29] Dennis 1725/1943, 278.
[30] Dennis 1720/1943, 181.

We have seen that Steele responded to the directly political attacks upon his nationality by constructing and circulating discursive accounts of himself as a sympathetic subject and by reconstructing his enemies as objects of ethnic and religious antipathy. In his recent book, Charles A. Knight argues that what he calls Steele's 'double vision' (Irish and English) shaped his 'politics of sympathy'.[31] Certainly, in episodes such as the debates over whether to execute or pardon peers who joined the Jacobite Rebellion in 1715, Steele was emphatically on the side of mercy. In *A Letter to a Member concerning the Condemn'd Lords* (1716), he wrote:

> I never talked of Mercy and Clemency, but for the Sake of my King and Country, in whose Behalf I dare to say, That to be afraid to forgive, is as low as to be afraid to punish; and that all noble Geniuses in the Art of Government have less owed their Safety to Punishment and Terror, than Grace and Magnanimity.[32]

It seems likely that the sympathy he extended to the Scots Lords in that episode was a factor in his appointment to the Committee for Sequestrations in the aftermath of the rebellion, as his reputation for clemency would have made him more acceptable in the north. But the stress here on pity as a characteristic of political genius had an ideological as much as a tactical import. In *The Englishman* No. 32 (17 December 1713), Steele uses the familiar analogy of the kingdom as 'a great Family' to stress that a ruler needs to treat his subjects with 'Love, Tenderness, and Compassion', without which his authority will soon decay.[33] In a slightly later issue, he returns to the familial analogy to amplify his critique of absolutist monarchy:

> To say, therefore, that the Nature of Government requires an absolute Submission in the whole governed Society, even to a Degree of total Ruin, when that shall seem fit to the governing Part, is just as if it should, with great Gravity be affirmed, That the Nature of Government requires, that the very End for which only it was instituted, should be frustrated, and wholly destroyed. [...]
> It is as if it should be said, That the Nature of a Guardianship requires, that the Children, for whose Good it was settled, must, without Limitation, submit, should a Guardian sell them to the Slavery of the Galleys.[34]

Steele's example of illegitimate, despotic governance uses the same figure of literal enslavement that he invoked in his defence of his conduct in the *Apology*, in which, as we saw, he characterised his Parliamentary tormentor as a brutalised former slave. Literal enslavement often appears in Steele's polemical writing as the horrific telos of the political domination he associates with absolutism, while an inclusive compassion is good government's virtuous antithesis. But in his *Apology*, as clearly as in the celebrated fable *Inkle and Yarico*, the invocation of slavery actually collapses that crucial opposition between the free trading nation and its tyrannical rivals. Feeling himself not only humiliated but literally cast out of the political nation by his expulsion from the House, Steele's rhetorical

---

[31] Knight 2009, 12.
[32] Steele 1716/1944, 415.
[33] Steele 1955, 129.
[34] *The Englishman* No. 22, of 23 September 1715, in Steele 1955, 337.

misrecognition of his accuser as a depraved denizen of the slave trade points to a fundamental contradiction in his beloved 'English constitution', identifying slavery as a delegitimating canker on the British body politic. The dark body and 'exotick' origins that he shares with his accuser, rhetorically at least, continually threaten expulsion from the idealised free Protestant nation into a condition of slavery whether material or political.

2.2 Play Making

Forced to reflect repeatedly upon his own and others' nationalities, it is unsurprising that Steele seems peculiarly sensitive to the shifting, contingent, and hierarchical nature of ethnic and national identities and affects. With multiple national affiliations and residencies, he was in pole position to originate literary techniques that manage difference and distance. While the flow of sympathy in sentimental drama (or novels) may help constitute and consolidate identities, it also blurs boundaries, whether one figures the sympathetic self as subjugated by feeling or aggressively appropriative of another's most intimate experience via identification. Given Steele's own always uncertain hold upon an 'English Constitution', it makes sense he should find the 'universal' appeal of virtuous sympathy—in which national, party, and even gender differences are putatively sunk—so compelling.

When he finally produced *The Conscious Lovers*, generally accepted as the fullest exemplum of the sentimental comedy, Steele had been theorising the genre for years. In *Tatler* No. 172, he argues against the recourse to the 'History of Princes and Persons who act in high Spheres', believing in 'the great Use (if any Body could hit it) to lay before the World such Adventures as befall Persons not exalted above the common Level'.[35] He rejected 'poetical justice' and preferred plays 'in which the persons are all of them laudable, [in which] their misfortunes arise from unguarded virtue than propensity to vice'.[36] The aim of the dramatist was to unite the audience in a sympathetic response to suffering virtue, the sign of such pity being tears.

Where was the model for such a dramaturgy? Steele identifies it in the practice of John Banks, author of oriental heroic plays before he started composing a run of she-tragedies with subjects drawn exclusively from British history in the 1680s. For Steele, the great virtue of Banks's plays was that they were tear pumps, as he remarks (not altogether admiringly) in *Tatler* No. 14:

> Yesterday we were entertain'd with the Tragedy of the *Earl of Essex*, in which there is not one good line, and yet a Play which was never seen without drawing Tears from some part of the Audience; a remarkable instance, that the Soul is not to be mov'd by Words, but

---

[35] Steele 1754, Vol. 3, 246.
[36] Steele 1712/1776, 171–172.

Things; for the Incidents in this Drama are laid together so happily, that the Spectator makes the Play by Himself, by the Force which Circumstance has upon his Imagination.[37]

Colley Cibber, also a contender for the title of first sentimental dramatist, makes precisely the same point about Banks. Exhorting would-be playwrights to remember the primacy of plot, Cibber invokes the example of Banks:

> There are three Plays of his, The Earl of Essex, Anna Bullen, and Mary Queen of Scots, which tho' they are all written in the most barren, barbarous Stile, that was ever able to keep the Stage, have all interested the Hearts of his Auditors. To what then could this Success be owing, but to the Intrinsik, and naked Value of the Tales he has simply told us? There is Something so happy in the Disposition of all his Fables; all his chief Characters are thrown into such natural Circumstances of Distress, that their Misery or Affliction, wants very little Assistance from the Ornaments of Stile, or Words to speak them. [...] At such a Time, the attentive Audience supplies from his own Heart, whatever the Poet's Language may fall short of, in Expression, and melts himself into every pang of Humanity, which the like Misfortunes in real life could have inspired.[38]

Banks's she-tragedies, several of which were suppressed in the 1680s, all held the stage through the eighteenth century.[39] Although recent commentators, such as Louise Marshall and Christine Gerrard, stress Elizabeth's continuing value as an emblem of proper Protestant rule in the plays produced under the Hanoverians, John Watkins argues that the Elizabeth depicted in Banks's drama is a repudiation of her status as a great sovereign, in that she is refigured as a suffering tragic heroine whose miseries are essentially private.[40] In an account consonant with other recent readings of pathetic tragedy, Watkins argues that Elizabeth's tragic suffering models the conflicted interiority of the emergent bourgeois subject.[41] Without contesting the centrality of class mediation in these texts, I want to suggest that Banks's tragedies were also successful over a period of decades in moving significant portions of audiences because his heroines were domestic in both senses—primarily concerned with private passions and equally importantly, characters in the national narrative. Although his female protagonists were royal, they were figures from a shared and not-too-distant British past, thus diminishing the distance from the audience who were simultaneously united in watching a common history unfold.[42] Banks was himself very emphatic about the importance of his choice of domestic subjects, writ-

---

[37] Steele 1754, Vol. 1, 85.
[38] Cibber 1740/1968, 190.
[39] In his *Memoirs of the Life of David Garrick, Esq.*, Thomas Davies echoes Steele's and Cibber's praise of Banks: 'The Tragedy of the Earl of Essex, by Banks, had lain long neglected, though no play had ever produced a stronger effect upon an audience: for though the language is a wretched compound of low phrase and bombast expression, and is indeed much below criticism; yet in the art of moving the passions Banks has no superior'. (Davies 1780, 294.)
[40] Gerrard 2002; Marshall 2008; Watkins 2002, 185–186.
[41] See Brown 1982.
[42] Mark Sabor Phillips has tracked this process in historiography from the mid-century. See Phillips 2000. For a wide-ranging discussion of the pleasures recollections of Mary Queen of Scots provided through the eighteenth century, see Lewis 2000.

ing in the 'Preface to Anna Bullen' (1682) that unlike those of his fellow dramatists, 'His *Heroes* all to *England* are confin'd', suggesting further that this should ensure the spectators' approbation: 'To your own *Fathers* sure you will be kind'.[43]

Although *The Unhappy Favorite* and *The Island Queens* were both suppressed by the Lord Chancellor in the 1680s, Banks insisted that his plays were innocent of parallels. While it is hard to see dramas in which a Protestant ruler executes a Catholic Stuart heir as entirely free of contemporary reference, the subsequent reception history of these texts suggests that they were valued for their ability to unify audiences by means of specifically British subjects and affects.

When Steele came to write *The Conscious Lovers* (possibly as early as 1713, although the play only appeared in 1722), he created a comic form which skirts tragedy. This allowed him to use figures 'not above the common level', closer to the audience than Banks's British queens, but capable of generating a similar pathos. In the rough notes which Steele drew upon for his preface, he writes that 'Addison told me I had a faculty of drawing Tears—and bid me compare the Places in Virgil wherein the most judicious Poet made his Hero weep',[44] and while he himself thought Bevil's refusal to fight in the fourth act the play's most important scene, audiences and critics were agreed that the recognition scene in which Indiana is reunited with her father was the affective climax of the drama.

The action of the play is focused on Indiana, long-lost daughter of the East Indies merchant Sealand. Rescued from a lecherous French captor, Indiana has come penniless to London, where she is under the protection of the virtuous hero, Bevil Jr. Bevil's father wishes his son to marry the wealthy Lucinda, beloved by Bevil's friend Myrtle. The forced marriage is averted by Sealand's recognition of Indiana, thus paving the way for her match with Bevil Jr. Recent critics of *The Conscious Lovers* have focused on the inter-related issues of the play's thematic reconciliation of monied and landed interests and the question of aesthetic legitimacy raised by the novel sentimental form in which the action is cast. Lisa Freeman has argued that Steele's project of inculcating 'good breeding' by means of an exemplary comedy was challenged by accusations that his new genre was an illegitimate hybrid whose curbing of humour embodied a threat to liberty.[45] Nicole Horejisi reverses Freeman's account of the play's positive vision of overseas trade by stressing Indiana's vulnerability to accidents contingent on East Indian trafficking.[46] Peter Hynes revisits the question of legitimacy by analysing how Terence's cultural authority as a classical progenitor of tender comedies was invoked to defend both Steele's text and his larger project of dramatic reform.[47]

None of these critics pays attention to Steele's position as a figure whose cultural authority was fractured by his own hybrid national status, particularly after 1714.

---

[43] Banks 1682, n.p.
[44] Quoted in Aitken 1889, Vol. 2, 277, N. 1.
[45] Freeman 2002, 193–234.
[46] Horejisi 2003.
[47] Hynes 2004.

This seems particularly surprising because the fusillade of attacks on the play by Dennis follow the practice of earlier pamphleteers in focusing on Steele's Irishness. In *A Defence of Sir Fopling Flutter, A Comedy Written by Sir George Etheridge* (1722), Dennis's argument about literary authority turns on nativism as much as genteel status: 'I shall only add, that I would advise for the future, all the fine Gentlemen, who travel to *London* from *Tipperary*, to allow us *Englishmen*, to know what we mean, when we speak our own Language'.[48] Dennis is particularly incensed by what he sees as Steele's violation of Thalia because of his conviction that comic excellence is a peculiarly national trait: 'the very Boast and Glory of the *British* Stage is Comedy, in which *Great Britain* excels any other Country: Nay, we can show more good and entertaining Comedies than all the rest of *Europe*'.[49] In attacking *The Conscious Lovers*, Dennis invokes two kinds of authority: classical poetics and insider knowledge of overseas trade. The first, extended critique of the play uses an Aristotelian standard of verisimilitude to indict the text as insufficiently probable in repeatedly failing to create plausible social representation:

> But now this whole Dramatick Performance seems to me to be built upon several things which have no Foundation, either in Probability, or in Reason, or in Nature. The Father of *Indiana*, whose Name is *Danvers*, and who was formerly an eminent Merchant at *Bristol*, upon his Arrival from the *Indies*, from whence he returns with a great Estate, carries on a very great Trade at *London*, unbeknownst to his Friends and Relations at *Bristol*, under the Name of *Sealand*. Now this Fiction, without which there would be no Comedy, nor anything call'd a Comedy, is not supported by Probability, Reason or Nature.[50]

Dennis queries the strategy of concealment, the implausibility of Sealand never sending for news of his missing wife, sister, and child from Bristol and in particular, the unlikelihood of a merchant returning

> from the *Indies* with a vast Estate, and the World should not know either what he is, or what he was when he went thither, especially when he traded to every Part of the Globe. Or was there ever any great Merchant of *London* whose Family and Original was not known to the Merchants of *Bristol*?[51]

For Dennis, the trading world is too transparent, secure, and well-networked to allow women to be taken prisoner and 'disappeared'. He calls Indiana's capture 'Pregnant with Absurdity' because ''Tis highly improbable, that an *East-Indies* Vessel, which had Force enough to venture without a Convoy, should be taken by a Privateer'.[52] He finds it ridiculous that Indiana's aunt could send no letters from France asking for help and insists that even were she unable to write, not only 'the whole *East*-India Company but all *London* would have known what was become of

---

[48] Dennis 1722/1943, 245.
[49] Dennis 1723/1943, 252.
[50] Dennis 1723/1943, 263.
[51] Dennis 1723/1943, 263.
[52] Dennis 1723/1943, 268.

the Ship, at a time when so many News-Writers contended which could furnish the Town with the freshest News'.[53]

Dennis has other complaints, about Indiana's dubious claims to modesty and Bevil's unbelievable filial piety, although his emphasis falls heavily on what he regards as a travesty of overseas trade. But the incidents from which Steele has constructed his action are not simply romance tropes recycled from Terence's *Andria*—they are reminiscent of events in his own family history. One of his aunts was born in South Asia and named Indiana; his first wife's brother was captured by French privateers while sailing from the West Indies and was killed. Steele's dead brother-in-law left a 'Negro woman' and numerous children, all of whom were manumitted on his death but received no inheritance, being reduced to indigence. Steele was able to bring his own marriage plot to a happy conclusion when the West Indian inheritance he gained from his first wife facilitated his marriage to a woman who brought him a small landed estate.

Anglo-Irish adventurism was not, it seems, hard to cloak in the tropes of romance or Terentian antecedent. The point was to bring the audience into collusion with this particular version of the trials and triumphs of 'the new and numerous Gentry' deplored by Dennis as he warned of Irish cultural corruption:

> The Sentiments in *The Conscious Lovers* are often frivolous, false, and absurd; the Dialogue is awkward, clumsy, and spiritless; the Diction affected, impure, and barbarous, and too often *Hibernian*. Who, that is concern'd for the Honour of his Country, can see without Indignation whole Crowds of his Countrymen assembled to hear a Parcel of *Teagues* talking *Tipperary* together, and applauding what they say?[54]

Steele is able to effect the triumph of what Danvers/Sealand also identifies in a famous speech as a new 'species of gentry, that have grown into the world this last century' by yoking affiliation to sentiment.[55] Cibber's commentary on the reception of *Cato* makes it clear that the expression and avowal of a feeling response to Addison's play in performance was mandated—to remain unmoved would be to mark oneself as not just undiscriminating but profoundly unpatriotic. Ten years later, *The Conscious Lovers* sought to exercise a similar power; the play's ostensible programme of elite reconciliation, extreme filio-piety, rakish reform, and exemplary benevolence was to be enforced by the spontaneous, communal response to Indiana's reunion with her father. Commentators through the eighteenth century bear out the claim that the play moved audiences: an early sonnet 'To Sir Richard Steele' (1726) remarks 'At Sealand's Feet to see his Daughter lie/ Each tender Heart o'erflows with Tears of Joy'.[56] Another commentator, writing several decades later, recalls a famous anecdote:

> We have already observed, that it is impossible to witness the tender scenes of this comedy without emotion; that is, no man who has experienced the delicate solicitudes of love and

---

[53] Dennis 1723/1943, 268.
[54] Dennis 1723/1943, 274.
[55] Steele 1722/1723, 62.
[56] Heywood 1726, 226, Lines 5–6.

affection, can do it. Sir Richard has told us, that when one of the players told M Wilks, that there was a general weeping for Indiana, he politely observed 'that he would not fight the worse for it'.[57]

The early panegyrics to Steele celebrating the play were equally emphatic about the its patriotic effect: 'The *British* Fair, thy finish'd Model shown, /By *Indiana's* Conduct set their own',[58] declaims one celebrant, 'What *Briton* now, will reckon Vertue dull?' asks another.[59] By weeping, the audience demonstrated their incorporation of and assent to the sentimental norms modelled on stage. No response could distinguish one weeping spectator from another except an indifference which would mark the viewer as uncivil—self-condemned to unfeeling isolation. Almost maddened by his sense of alienation from this community of taste and feeling, Dennis proclaimed wildly in his *Remarks on a Play call'd The Conscious Lovers* that 'I am as to this Matter, in a State of Nature with these Persons.'[60] In a deliberate, direct riposte to the ethnic aspersions which follow, Benjamin Victor praised Steele's hybridity: 'the greatest Panegyrick upon you, is the unprejudic'd and bare Truth of your Character, the Fire of Youth, with the Sedateness of a Senator, and the modern Gayety of an *English* Gentleman, with the Noble Solidity of an Ancient *Briton*'.[61] Refusing the ethnocentric singularity of Dennis's definitions of comedy and national identity, Victor celebrates Steele's personal combination of opposites as a model of contemporary British manhood. The value of the sentimental drama was equivalent, as its novel union of dramatic elements succeeded in erasing ethnic as well as sectarian, status, or party differences in a temporary community of proper feeling.

In the decades following the great success of *The Conscious Lovers,* other writers adopted pathetic and sentimental scenarios in pursuit of agendas which reiterated but also extended beyond Steele's conventional unionist, latitudinarian Anglican and Whig apologetics. Deist sympathisers such as John Hughes, Aaron Hill, and James Thomson wrote highly pathetic philo-Islamic plays which implicitly supported universal toleration; George Coleman the Elder adapted Voltaire's sentimental *L'Ecossaise* to rebuke contemporary Scotophobia and encourage inter-union harmony. Plays about cruelly treated Indians reiterated the black legend of Spanish conquest in America and set up an implicit contrast with British colonial policy. As abolition became a heated topic of cultural and political debate from the 1760s on, Southerne's *Oroonoko; or, The Royal Slave* was repeatedly revised to excise its comedy, heighten its pathos, and drive home an abolitionist message. Just how successful anti-slavery drama was in confirming rather than undermining African humanity is still an open question, but there is no doubt that contemporary commentators believed it to be an effective weapon, remarking of Coleman's sentimental *Inkle and Yarico* that it was 'as capable of *writing* a petition for the abolition of

---

[57] Ashford 1768, Vol. 6, xiv.
[58] Anonymous 1726, 68.
[59] Mitchell 1729, 257.
[60] Dennis 1723/1943, 257.
[61] Victor 1722, 29.

the slave-trade as any of those associated bodies who have taken so much pains for that laudable purpose'.[62] It is equally clear that pathetic and sentimental drama worked persistently to confirm spectators in their own sense of national superiority, not least their possession of that most vital of Enlightenment virtues, humane feeling. What seems more surprising is that they convinced others of it too: in a letter about *Oroonoko* sent by a French traveller to a friend back home, Jean Bernard Le Blanc commented: 'The author has painted the strongest of all virtues in it, with the strongest and most moving strokes; and let us say to the honour of the English, that which is the peculiar characteristic of their nation, humanity'.[63] If sentimental drama did not succeed in abolishing the slave trade, it certainly assisted in the construction of a national imaginary in which humane feeling assumed a central role.

References

Aitken, George A. 1889. *The life of Richard Steele,* 2 vols. London: Wm. Isbister.
Anonymous. 1726. To Sir Richard Steele, on his comedy, The Conscious Lovers. In *Miscellaneous poems by several hands*, 66–70. London: Printed by D. Lewis.
Ashford, J. 1768. *The theatre: or, select works of the British dramatick poets*, 12 vols. Edinburgh: Printed by Martin and Wotherspoon.
Banks, John. 1682. *Preface to a new play called Anna Bullen*. London: Printed for Allan Banks.
Bernbaum, Ernest. 1958. *The drama of sensibility: A sketch of the history of English sentimental comedy and domestic tragedy, 1696–1780*. Gloucester: P. Smith.
Boulukos, George. 2009. *The grateful slave: The emergence of race in eighteenth-century Britain and America*. Cambridge: Cambridge University Press.
Brown, Laura. 1982. The defenceless woman and the development of English drama. *Studies in English Literature, 1500–1900* 22(3): 429–443.
Carey, Brycchan. 2005. *British abolitionism and the rhetoric of sensibility: Writing, sentiment and slavery, 1760–1807*. New York: Palgrave MacMillan.
Cibber, Colley. 1740/1968. *An apology for the life of Colley Cibber with an historical view of the stage during his own time written by himself*, ed. B. R. S. Fone. Ann Arbor: University of Michigan Press.
Coleman, Deirdre. 2005. *Romantic colonization and British anti-slavery*. Cambridge: Cambridge University Press.
Colley, Linda. 1994. *Britons: Forging the nation, 1707–1837*. New Haven: Yale University Press.
Davies, Thomas. 1780. *Memoirs of the life of David Garrick, Esq*. London: Printed for the Author.
Defoe, Daniel. 1713. *The honour and prerogative of the Queen's majesty vindicated and defended against the unexampled insolence of the author of the guardian: In a letter from a country Whig to Mr Steele*. London: Printed for John Morphew.
Dennis, John. 1720/1943. The character and conduct of Sir John Edgar, call'd by himself sole monarch of the stage in Drury-Lane; and his three doughty governors. In *The critical works of John Dennis*, 2 vols., vol. 2, ed. Edward Niles Hooker, 181–199. Baltimore: The Johns Hopkins University Press.
Dennis, John. 1722/1943. A defence of Sir Fopling Flutter, a comedy by Sir George Etheridge. In *The critical works of John Dennis*, 2 vols., vol. 2, ed. Edward Niles Hooker, 241–250. Baltimore: The Johns Hopkins University Press.

---

[62] *The Theatrical Register*. 1788 (York), 12–13, cited in Worrall 2007, 1.
[63] Le Blanc 1747, Letter LVI, Vol. 2, 58.

Dennis, John. 1723/1943. Remarks on a play, Call'd The Conscious Lovers, a comedy. In *The critical works of John Dennis*, 2 vols., vol. 2, ed. Edward Niles Hooker, 251–274. Baltimore: The Johns Hopkins University Press.

Dennis, John. 1725/1943. The causes of the decay and defects of dramatick poetry, and of the degeneracy of the publick taste. In *The critical works of John Dennis*, 2 vols., vol. 2, ed. Edward Niles Hooker, 275–299. Baltimore: The Johns Hopkins University Press.

Ellis, Markman. 1996. *The politics of sensibility: Race, gender and commerce in the sentimental novel*. Cambridge: Cambridge University Press.

Festa, Lynn. 2006. *Sentimental figures of empire in eighteenth-century Britain and France*. Baltimore: The Johns Hopkins University Press.

Freeman, Lisa A. 2002. *Character's theater: Genre and identity on the eighteenth-century English stage*. Philadelphia: University of Pennsylvania Press.

Gerrard, Christine. 2002. *The patriot opposition to Walpole: Politics, party and national myth 1725–1742*. Oxford: The Clarendon Press.

Gottlieb, Evan. 2007. *Feeling British: Sympathy and national identity in Scottish and English writing, 1707–1832*. Lewisburg: Bucknell University Press.

Heywood, James. 1726. To Sir Richard Steele, on his comedy, call'd The Conscious Lovers. In *Letters and poems on several subjects*, 226. London: Printed for W. Meadowes, T. Worral.

Horejisi, Nicole. 2003. (Re)Valuing the 'Foreign-Trinket': Sentimentalizing the language of economics in Steele's *Conscious Lovers*. *Restoration and Eighteenth-Century Theater Research* 18(2): 11–36.

Hynes, Peter. 2004. Richard Steele and the genealogy of sentimental drama: A reading of *The Conscious Lovers*. *Restoration and Eighteenth-Century Theater Research* 40(2): 142–166.

Kaul, Suvir. 2000. *Poems of nation, anthems of empire: English verse in the long eighteenth century*. Charlottesville: University Press of Virginia.

Kidd, Colin. 1999. *British identities before nationalism: Ethnicity and nationhood in the Atlantic world 1600–1800*. Cambridge: Cambridge University Press.

Knight, Charles A. 2009. *A political biography of Richard Steele*. London: Pickering and Chatto.

Lacy, John. 1714. *The ecclesiastical and political history of Whig Land, of late years*. London: Printed for John Morphew.

Le Blanc, Jean Bernard. 1747. *Letters on the English and French nations*, 2 vols. London: Printed for J. Brindley, R. Francklin, C. Davis, J. Hodges.

Lewis, Jayne. 2000. 'The *sorrow* of seeing the Queen': Mary Queen of Scots and the British history of sensibility, 1707–1789. In *Passionate encounters in a time of sensibility*, ed. Maximillian E. Novak and Anne Mellor, 193–220. London/Cranbury: University of Delaware Press/Associated University Presses.

Manley, Delariviere. 1720. *Secret memoirs and manners of several persons of quality of both sexes. From the new Atalantis, an Island in the Mediterranean*, 6th ed, 4 vols. London: John Morphew.

Marshall, David. 1986. *The figure of theatre; Shaftesbury, Defoe, Adam Smith, and George Eliot*. New York: Columbia University Press.

Marshall, Louise H. 2008. *National myth, imperial fantasy: Representations of British identity in the early eighteenth century*. London: Palgrave.

Mitchell, Joseph. 1729. To Richard Steele on the successful representation of his excellent comedy call'd, The Conscious Lovers. In *Poems on several occasions*, 2 vols., vol. 2, 255–258. London: Printed by L. Gilliver.

O'Quinn, Daniel. 2005. *Staging governance: Theatrical imperialism in London, 1770–1800*. Baltimore: The Johns Hopkins University Press.

Orr, Bridget. 2001. *Empire on the English stage, 1660–1714*. Cambridge: Cambridge University Press.

Phillips, Mark Sabor. 2000. *Society and sentiment: The genres of historical writing in Britain 1740–1820*. Princeton: Princeton University Press.

Roach, Joseph. 1996. *Cities of the dead: Circum-Atlantic performance*. New York: Columbia University Press.

Shields, Juliet. 2010. *Sentimental literature and Anglo-Scottish identity, 1745–1820*. Cambridge: Cambridge University Press.

Steele, Richard. 1712/1776. The Spectator No. 290. Friday, February 1. In *The Spectator. Volume the Fourth*, 169–172. Edinburgh.
Steele, Richard. 1714/1944. An apology for himself and his writings. In *Tracts and pamphlets by Richard Steele*, ed. Rae Blanchard, 275–346. Baltimore: The Johns Hopkins University Press.
Steele, Richard. 1716/1944. A letter to a member, etc. concerning the Condemn'd Lords, in vindication of gentlemen Calumniated in the St. James's post of Friday March the 2nd. In *Tracts and pamphlets by Richard Steele*, ed. Rae Blanchard, 239–254. Baltimore: The Johns Hopkins University Press.
Steele, Richard. 1722/1723. *The conscious lovers. A comedy as it is acted at the theatre royal in Drury-Lane*. London: J. Tonson.
Steele, Richard. 1754. *The Tatler: The Lucubrations of Isaac Bickerstaffe, Esq*, 4 vols. London: Printed for H. Lintott et al.
Steele, Richard. 1955. *The Englishman: A political journal by Richard Steele*, ed. Rae Blanchard. Oxford: The Clarendon Press.
Steele, Richard. 1959. *Richard Steele's periodical journalism 1714–1716*, ed. Rae Blanchard. Oxford: Clarendon Press.
Swaminathan, Srividhya. 2009. *Debating the slave trade: Rhetoric of British national identity, 1759–1815*. Farnham/Burlington: Ashgate.
Swift, Jonathan. 1714. *The Publick spirit of the Whigs*. London: Printed for T. Cole.
Sypher, Wylie. 1969. *Guinea's captive kings: British anti-slavery literature of the XVIIIth century*. New York: Octagon Books.
Victor, Benjamin. 1722. *An epistle to Richard Steele, on his play call'd The Conscious Lovers*, 2nd ed. London: Printed for W. Chetwood, S. Chapman, J. Stagg, J. Brotherton, Th. Edlin.
Wagstaffe, William. 1714. *A letter from the facetious Doctor Andrew Tripe at Bath to the venerable Nestor Ironside*. London: printed by B. Waters.
Watkins, John. 2002. *Representing Elizabeth in Stuart England*. Cambridge: Cambridge University Press.
Worrall, David. 2007. *Harlequin empire: Race, ethnicity and the drama of the popular enlightenment*. London: Pickering & Chatto.

Chapter 3
Rochester's Libertine Poetry as Philosophical Education

Brandon Chua and Justin Clemens

Abstract Whatever the interminable discussions regarding the true meaning and import of Hobbes' political philosophy, one key question it poses to post-Restoration society is this: if the state is indeed an artificial man that requires the sacrifice of a natural portion of our being to enter, what—beyond pure violence and doctrine to shape the drive to self-preservation—is capable of holding it together? If bodies are themselves composites, refashioned and maintained by the material quest for pleasure and vainglory, what parts must be sacrificed to polity, and what happens to those parts necessarily excised in becoming part of a polity? These questions precipitate a crisis in the thought, experience, and acts of individuals. Libertinism is one of the attempted solutions to this crisis. It is, moreover, a paradoxical solution that, in explicitly exacerbating the aporias of materialism—that is, in literally digging its own grave—offers new possibilities for embodied action that are taken up by subsequent thinkers, anticipating (if in a wittier and less prolix fashion) the writings around 'sensibility' from Sterne to Sade. This article argues that the acts and writings of John Wilmot (1647–1680), the Earl of Rochester, exemplary libertine, poet and courtier, show him to be a crucial negative precursor for the theories of 'sensibility' that dominated the following century.

B. Chua (✉)
Centre for the History of Emotions, The University of Queensland,
4072 St Lucia, QLD, Australia
e-mail: b.chua1@uq.edu.au

J. Clemens
School of Culture and Communication, University of Melbourne,
John Medley Building, 3010 Parkville, VIC, Australia
e-mail: jclemens@unimelb.edu.au

3.1 John Wilmot, Earl of Rochester, a Precursor of Sensibility?

In this essay, we argue that John Wilmot, Earl of Rochester, is an unheralded harbinger of a sensibility he would himself have despised. This phenomenon has been missed by critics of the period because they misconceive the relationship between the philosophy of Thomas Hobbes and its singular embodiment by Rochester. Through a close reading of Rochester's *A Ramble in St. James's Park*,[1] we suggest that Rochester's engagement with Hobbes be read as a critical application of Hobbes's political theory, an engagement that is unable to be properly accounted for in the critical paradigms placing Rochester as a misguided and overly-enthusiastic disciple of Hobbes on the one hand, or a rebellious son seeking to overthrow his philosophical father on the other. While critics interested in the Rochester/Hobbes relation have traditionally focused on the overtly metaphysical *Satyr on Reason and Mankind*, we propose that a closer analysis of Rochester's rambling exploration of the socially dynamic spaces of the newly refurbished St. James's Park enables a fuller account of Rochester's reading of Hobbes's political theory to emerge. If, as Alexander Cook asserts in this volume, the development of a new mode of sociability predicated around the cult of sensibility can be regarded in one sense as a rejection of a 'Hobbesian world of calculation and egoism', we seek to complicate the terms of this rejection. Through a re-examination of the discursive relationship between Hobbes and Rochester, we retrace a winding historical path to sensibility forged from within what seems on the surface a near-total identification on the part of the poet-disciple with his philosophical master. Such an account, we suggest, requires that we revise traditional understandings of philosophical influence and the formal divisions separating philosophy and literature. In reformulating our understandings of 'philosophy', 'the literary', and 'education', we provide a reconsideration of Rochester's libertine ramble as a committed critique of Hobbes's account of political and social obligation. In doing so, we also suggest that, far from sensibility simply emerging as a kind of metaphysico-medico-literary reaction *against* Hobbes, it requires as one of its conditions the exacerbation of a kind of conceptual abscess introduced within Hobbesianism by Hobbesians themselves.

3.1.1 The 'Influence' of Thomas Hobbes on Atheist Libertinism

That Rochester was a devotee of Hobbes is a critical commonplace. Since Robert Parsons, Chaplain to Lady Rochester and one of the witnesses to the Earl's infamous deathbed conversion, wrote of Rochester's repudiation of 'that absurd and foolish Philosophy, which the world so much admired, propagated by the late Mr Hobbs, and

---

[1] Wilmot 1999.

others, [that] had undone him, and many more of the best parts of the Nation,'[2] critics have enthusiastically speculated on the nature of this relationship. Since at least 1903, when William Courthorpe argued that Rochester was entirely indebted to the Hobbesian philosophy, the speculations have intensified.[3]

Scholarly attention on Rochester's Hobbesian moment has typically focused on Rochester's engagement with Hobbes's redefinition of motion, as well as his materialist re-conception of the human will. The object of several of these critical explications on Hobbes's philosophical influence has been Rochester's most overtly philosophical poem featuring an ironic account of the contemplative life, the *Satyr on Reason and Mankind*.[4] In 1973, K. E. Robinson issued a useful summary of the key critical positions mapping the terms of Rochester's debt to Hobbes:

> either they [i.e. critics] believe with Courthorpe that Rochester 'puts forward his principles, moral and religious, such as they are, with living force and pungency, showing in every line how eagerly he has imbibed the opinions of Hobbes'; or they take Pinto's line that having 'started as a wholehearted disciple of Thomas Hobbes' he shows himself in the Satyr to be moving towards a 'bitterly ironic commentary on the mechanistic conception of humanity which was the logical outcome of the new science'; or they align themselves with Fujimura's view that Rochester owes a debt to Hobbes only in so far as he was a formative influence upon what was ultimately a much more pessimistic philosophy.[5]

Rochester, in this critical summary, is either an eager disciple parroting back Hobbesian tenets in the less philosophically rigorous medium of poetic satire, or he is the ironic commentator on trendy philosophical precepts, using the s*prezzatura* of an aristocratic *cortegiano* to deride or affirm the reductionist tendencies of the new, fashionable materialism. Robinson, quite brilliantly, reads this critical dissensus as the product of Rochester's own poetic program: the latter's 'irony requires such different readings to exist side by side'.[6] This irony is itself historicised, with Robinson reading Rochester's flagrant omission of the exceptional, unifying power of the sovereign as the basis of human sociability—or, rather, his reduction of the monarch himself to an entirely material entity—as a commentary on the new political order: what the Restoration presents is not a properly restored sovereignty, but rather disavowed and hypocritical disorder.

Since Robinson's summary of the critical tradition, readings of Rochester's poetry have largely moved away from Hobbesian metaphysics—sometimes abandoning Hobbes's influence altogether—and towards a more contextualising account

---

[2] Parsons 1681, 13. For a good selection of the historical criticism from Parsons and Gilbert Burnet to Edmund Gosse and Walter Raleigh, see Farley-Hills 1972.

[3] Courthorpe 1903, 465.

[4] The critical commentary on Rochester's *Satyr* is voluminous, in stark contrast with the rest of his poetic output. See, for instance, Moore 1943, 393–401; Fujimura 1958, 582; Berman 1964, 364–365; Knight 1970, 254–260; Johnson 1975, 365–374; Robinson 1973, 108; Cousins 1984, 429–439; and Russell 1986, 246.

[5] Robinson 1973, 108.

[6] Robinson 1973, 109.

of Rochester's relationship with a more broadly defined Restoration culture.[7] Locating an emerging, modern social consciousness in the period after the Restoration, recent readings of Rochester's poetry have emphasised his position in a rapidly shifting social order marked by mobility and flux.[8] Such work has moved critical discussion away from abstract philosophical concerns over matter and motion and towards Rochester's implied critique of emerging forms of sociability enabled by the gradual destabilisation of traditional social hierarchies. Class mobility, an emerging market economy, and the increasing devolution of aristocratic status comprise the contextual terms of recent critical re-readings of Rochester's poetry—re-readings that have had little to say about the influence of the 'absurd late philosophy' propagated by one, Mr Hobbes. Insofar as critics have been interested in Rochester's debt to Hobbes, the interest has remained largely in the realm of metaphysics, with Rochester's attack on the rational faculties of man taking centre stage. Meanwhile, the critical move towards reading an emerging social consciousness has had little time for questioning the Hobbesian dimensions of Rochester's critique of sociability and obligation. We show how these two positions can be united through a reconsideration of Rochester's reading of Hobbes. We maintain, with the priggish Robert Parsons, that Rochester is indeed Hobbes's disciple. But we also show how he is a devoted Hobbesian in a social and political context. The critical attention on Rochester's attempt to define Reason or Motion after the Hobbesian moment misses something crucial about Rochester's reading of Hobbes. For Rochester cannot be considered a philosopher, under any description offered by the aforementioned commentators. As such, the problem of 'influence' is essentially disavowed: on the one hand, sources and allusions are carefully and dutifully investigated; on the other, the source- and allusion-hunting are misconceived insofar as the notion of influence itself is not adequately questioned. What earlier criticism has properly recognised is that Rochester is indeed Hobbes's disciple, that he read Hobbes extremely closely, and that he absorbed much of the Hobbesian metaphysics. However, what these critics miss in treating the relationship as one of simply textual and argumentative influence is precisely the fact that Rochester decided to *live out* the implications of the Hobbesian political compact.

Given, then, the absolute scholarly consensus that Hobbes was indeed an 'influence' on Rochester and the concomitant dissension regarding the specifics of this influence, has any new information come to light that might legitimate yet another article

---

[7] In the same year that Robinson's summary appeared, Jeremy Treglown published an article on Rochester's debt to English sources, leaving Hobbes altogether aside, while locating Rochester in an older English poetic tradition. See Treglown 1973, 42–48.

[8] See Chernaik 1995. Chernaik's important contribution to the Hobbes/Rochester relationship pays sharp attention to the cultural work performed by the Restoration wits' appropriation of Hobbesian philosophy in a society undergoing a destabilising shift from a culture of status to one of contract. Chernaik, however, leaves no room for the possibility that Rochester was engaging Hobbes in anything but a haphazard, careless, and inconsistent fashion. Other contextualising works, in their emphasis on the performative dimension of a Hobbes-inspired libertine lifestyle, similarly fail to elaborate on a more specific engagement with Hobbes's political philosophy. See Webster 2005; Turner 2002; Combe, 1998.

on this relationship? It has not. What, however, has inspired the current essay is a conviction that the forms of attention to Rochester's Hobbesian debt have hitherto been shaped by sets of assumptions—assumptions about the status of Rochester's allusions, the import of genre, the relation between literature and philosophy, and the role of ethics—that vitiate the possibility of a properly *philosophical* articulation between the two.

3.1.1.1 Excursus on Method: Philosophy as Education

We propose a quite different mode of interpretation to the tradition here. Our thesis is the following: we wish to show how the life and work of Rochester, the Restoration libertine, cannot be understood outside of his struggle with the thought of Hobbes, a struggle which entailed embodying the consequences of Hobbes' doctrines regarding matter, motion, reason, and sovereign power. In doing so, Rochester takes Hobbes' philosophy to its limit, exposing as he does so its central unthought elements. He thus in no way *resembles* his master, although, as we shall see, he is indeed one of the radical outcomes of Hobbes; moreover, in this, he also exits philosophy in favour of poetry; by this move, he becomes philosophically important in a way that can only be misrecognised by both philosophers and poets—and this is part of the point.

The traditional critical emphasis on the largely speculative nature of Rochester's debt to Hobbes is also the symptom of critics misreading the legacy of Hobbes in Restoration culture. In his recent study of Hobbes's reception in the later seventeenth-century, Jon Parkin argues that scholarship has largely taken Hobbes's contemporary opponents at their word, accepting uncritically their characterisations of Hobbes as a conservative, totalitarian pessimist whose ideas were a 'bizarre aberration in seventeenth-century intellectual history'.[9] Critics of Restoration culture have largely taken for granted the almost wholly negative impact of Hobbes's political ideas among his contemporaries. For Parkin, however, the repeated disavowals of Hobbes's philosophy by the architects of the Restoration settlement such as Clarendon, members of the Cavalier parliament, and the high Anglican establishment, instead reveal subtle attempts to assimilate Hobbes's theories of sovereign power into a normative political vocabulary. If we read the anxious repudiations of Hobbes as acknowledgments of his centrality in considerations of authority and obligation, then, we find an ongoing engagement with his political philosophy that belies the anxious affirmations of his marginality. If we reposition Hobbes as a central political figure in the Restoration, whose political theory was publicly rejected precisely because, as Parkin argues, it offered contemporaries tantalising solutions to hitherto intractable political problems, then we can re-read Rochester's engagement with Hobbes in terms of a political debate, rather than as a performance of subversive tenets geared towards a petulant aristocratic desire to shock orthodox

---

[9] Parkin 2007, 1.

sensibilities.[10] Victoria Kahn has similarly observed the marginal position occupied by Hobbes in critical accounts of the rise of a distinctly modern form of political economy in the late seventeenth century.[11] According to Kahn's account, Hobbes's misanthropic support for absolutism often compromises his place in the history of the proto-liberal subject. While Locke's possessive individualism is seen as crucial to the general culture of the period, Hobbes is associated 'with a cultural moment that was passing away rather than with one that was emerging'.[12] Through re-reading Rochester's dialogue in terms of political theory, we seek to build on the recent critical reconsiderations of Hobbes's place in the history of modern subjectivity. We seek to show, in particular, how Rochester's Hobbesian reading of sociability and affect creates a space for the immediate subsequent theories of *sensibilité* to flower.

This is not to say that we do not rely heavily on the commentators of a Hobbesian Rochester to date. On the contrary, we can immediately agree that: (1) Hobbes is a primary source for Rochester, at the level of allusion and argument; (2) that the major features of Rochester's indebtedness include Hobbes's epistemology, his materialism, the analysis of fear as the foundation of human community, and the concomitant inference that value-judgements are nominalistic; (3) that Rochester is nonetheless not Hobbes, and indeed exacerbates the latter at certain points, and criticises him at others (regarding the role of fear, the reduction of the monarch to material, etc.). What we add is a new account of the relationship between Hobbes and Rochester, which at once explains details in Rochester's work which have as yet gone under-remarked or misunderstood, and which can account for many of the points of critical dispute.

Our methodological point has to do with reconceiving the relation between philosophy, its genres, and education. We have been alerted to the absolute centrality of this theme within philosophy itself by A.J. Bartlett.[13] Our initial principles can be summarised telegraphically in the following points:

1. Philosophy is itself integrally educational.
2. There is therefore no such thing as a 'philosophy of education'.
3. Philosophy does not educate by means of doctrine or propositions, but operates upon existing 'state' practices of education.
4. Philosophy is therefore a discourse directed towards *re*-education.
5. Philosophy is an inventive practice of re-education through 'truths'.

We take these points as part of the self-definition of philosophy itself. Indeed, if one takes philosophy's re-educational function as primary, then it is immediately

---

[10] For a critique of Rochester scholarship that takes issue with the critical tendency to dismiss the seriousness of Rochester's works, seeing them as the idle amusements of a spoiled, attention-seeking rake, see Combe 1998. While we agree with Combe on the need to read the poetry and performance as contributions to political argument and theory, Combe, however, doesn't extensively address Rochester's engagement with Hobbes's political theory of sovereignty.

[11] Kahn 2004, 21–24.

[12] Kahn 2004, 23.

[13] See Bartlett 2011.

possible to draw real links between philosophers whose thoughts are otherwise irremediably disjunct.

Such a theory demands that the interpretation of philosophers be at once thoroughly situated—it is only by examining philosophers with respect to their particular time and place that we are able properly to delineate the problems to which they were responding ('history of philosophy')—and thoroughly creative. That is, it is in circumscribing their problems that we can also determine their inventiveness. But this also generates several further paradoxes. First, the very relation between 'problem' and 'solution' is itself philosophically variable, and sometimes even its inventive conception of a problem is itself the key element of a philosophy. Second, it is not simply possible to 'read backwards' what is philosophy and what is not. To give only a single example, which will become important in our reading of Rochester: the mode of presentation is integral to the philosophy itself, and sometimes that philosophy will express itself 'poetically'. Just because somebody writes a learned treatise on logic does not necessarily make that person a philosopher; just because somebody writes a poem does not necessarily make that person not a philosopher: Parmenides, Lucretius, Dante. Third, it makes *ethics* the key to philosophy, that is, a practice without a model.

What this means in this context is that Rochester's relationship to Hobbes is at once absolute and unique. Rochester's life is an experiment on the basis of Hobbes's new materialism: simply put, the question Rochester poses is, *if Hobbes is right, then how should I live*? This question is operative at every level of Rochester's life, from his admiration of Hobbes in his courtly performances, to the satires covertly circulated among intimates. In this living-out of a philosopher's program, however, dangerous questions arise in its course, such as: *how is it that the pleasure that ought to secure the stability of the body destroys it*? and *how is it that the legitimate pursuit of pleasures generates evaluations contrary to the grounds of the pleasures themselves*? If these questions have an impeccable philosophical pedigree, the answers that can be given them will necessarily have to be new, post-Hobbesian answers.

We would also like to note that this situation may be phrased in a number of apparently very different vocabularies. Harold Bloom's theory of the 'anxiety of influence' is also integrally a (philosophical) theory of education-by-the-literary.[14] It is therefore quite surprising how often this integral aspect of Bloom's theory is missed or underplayed in the secondary literature. For Bloom, the 'books and schools of the ages', are so precisely because they do not give propositional lessons, informing their subjects with useful facts and methodologies, but because they, in their very substance, create *new* phrasings and affects from the matter of their predecessors. 'Strong poets' are enigmatic and multiple for Bloom, presenting their readers with singular takes on the problems of surviving death; moreover, they often do so by presenting radical 'scenes of instruction' that exceed the powers of any authorised guides to properly transmit them. Literature, just as we have been saying of philosophy, is therefore a self-authorising discourse for Bloom. Our point of

---

[14] See, inter alia, Bloom 1973, 1975a, b, 1982.

departure from Bloom's account is dependent upon two major factors. First, if Bloom is right to erase inherited disciplinary definitions regarding 'literature', 'philosophy', 'psychoanalysis', etc., he is wrong to subsume everything under the heading of the literary. Philosophy always makes a claim on 'the real' (however that is conceived) that is supplementary to the sorts of literariness in which Bloom is interested. Second, the six-fold set of tropes with which Bloom identifies his own 'revisionary ratios' are themselves too narrow to account for the generality posited by the philosophies with which we wish to engage. As we will see with Rochester and Hobbes, to reduce this relationship to a struggle for priority as if the truth-effects it generates are secondary is to miss certain crucial aspects of their work.

What all of these theories nevertheless share is, to reiterate, a very traditional view of education. As Jacques Lacan notes in his notorious essay 'Kant avec Sade':

> I, on the contrary, maintain that the Sadean bedroom is of the same stature as those places from which the schools of ancient philosophy borrowed their names: Academy, Lyceum, and Stoa. Here as there, one paves the way for science by rectifying one's ethical position.[15]

To Lacan's list, we will add—pertinent in a context in which Rochester himself translated fragments of Lucretius—the Epicurean Garden. For the Epicurean materialists, of course, chance rules the universe, which is nothing but aggregates of atoms, void, and swerve, and in which the gods are absent and careless of our fates. But the Epicurean withdrawal into moderated enclosed friendship is impossible for Rochester, for whom the court and town at night are his purview—and for whom the theatre is a paradigm of real life.

3.2 A Ramble in St. James's Park

A Ramble in St. James's Park is one of the most important poems in Rochester's slim *oeuvre*. Along with poems of the same era (approximately 1672–1675), such as *Upon Nothing* and the extraordinary *A Satyr against Reason and Mankind*, the *Ramble* is widely considered among Rochester's signature productions, joining the other licentious *tours-de-force* that are *The Imperfect Enjoyment* and *Signior Dildo*. What we wish to show is that this poem—usually, as we have seen, read as a kind of sexualised versification of a mixed materialist mode—is, rather, a kind of thought experiment that has to be understood as something quite other than a transcription of philosophical attitudes into ribald metres. To put it another way, this poem is one of a number of thought-experiments by Rochester, which poses the question: *If Thomas Hobbes is right about materiality and sovereignty, then what are the consequences for individual action?*

The versified account of a ramble through one of London's most prominent public sites draws attention to the relationship between social intercourse and base

---

[15] Lacan 1966, 645.

appetite, between public convention and private desires, through the poem's persistent usage of sacred imagery that it proceeds to deflate and debase. As the narrator arrives at St. James's Park, a prominent site located near Whitehall Palace open to the public and often frequented by Charles II and his circle of 'wits', he launches into an etiological account of the park's facilitation of erotic intrigue, thus offering sexual instinct as the origins of the public, 'hallowed walks' of St. James's Park.[16] The narrative assumes the generic conventions of royalist historiography and its sacralising tendencies, but deploys them towards a distinctly de-sacralising end: 'But though St. James has th'honour on't,/'Tis consecrate to prick and cunt'.[17] Playing on the disjunction between the park's present reputation for amorous intrigue and its religious epithet, the narrator renders the category of the sacred a mere product of inflated rhetoric by audaciously consecrating the sexual organs of the park's visitors.

Rochester's account of the promiscuous spaces of the park can be read as a mocking response to Edmund Waller's poetic description of St. James's Park, a description that accounts for the park's pastoral beauty by way of a celebration of the entire kingdom as a virtuous community governed by a temperate and just sovereign. In reading the elaborate renovation of the park after the fires of 1666 as a royal gift bestowed on an elated and grateful public, Waller's poem hails St. James's Park as a great civic accomplishment borne out of what Kevin Sharpe has described as the 'reciprocal love' joining ruler and subjects, articulating a royalist vision of social and political obligation based upon traditional notions of the monarch's divine right to govern and the subject's virtuous duty to obey.[18]

Waller associates the renovation of St. James's Park after the fires of 1666 with the mythic pioneers of civilised societies, Orpheus and Amphion.[19] Charles II is placed in this line of civilising figures, as a bringer of order to chaos, returning prosperity and civility to a community racked with 'popular rage'.[20] For Waller, the newly reconstructed park is a symbol of the triumph of monarchy over popular, divided rule, an affirmation of the legitimacy of the Stuart government, a legitimacy grounded in unbroken tradition and divine favour. Waller's pastoral celebration of the park's public yet intimate spaces renders the site as a space of virtuous contemplation frequented by a Philosopher-King, whose authority is exerted through his displays of selfless concern for the public interest and justice. Waller's panegyric activates an understanding of government as a structure of discipline and constraint based on the virtue of temperance, with the park allowing Waller's ideal monarch to exhibit his subordination of his 'private passions' to the 'public cares' of the community. An idealised picture of government is presented here, with the self-disciplined monarch

---

[16] On the popularity of St. James's Park among court wits and its reputation for amorous intrigue, see Narain 2005, 559.

[17] Wilmot 1999, Lines 9–10.

[18] Sharpe 1987, 168.

[19] For a brilliant reading of the ambiguous nature of poetic references to these two mythic *figures* of civilisation in early-modern poetry, see Greene 1982, 233–241.

[20] Waller 2001, 500–503, Line 98.

governing a tempered community. The hallowed walks of St. James's Park are thus presented as an analogue of Charles's well-governed kingdoms, with ruler and subjects united in the pursuit of the common good.

Rochester's redescription of St. James's Park turns Waller's microcosm of well-governed order into an account of a drunken, rambling pursuit of private pleasure, subversively challenging Waller's account of political obligation while raising troubling questions on the foundations of political community. Rochester's narrator, embarking on his ramble from drinking hole to park, proceeds to deploy the rhetoric of mythic history not to praise the Restoration monarchy, but to document the origins of the park's clandestine facilitation of sexual coupling—a narrative featuring an act of sexual onanism performed by the Ancient Picts, who, like the narrator, turn out to be early victims of 'jilting':

> For they [the strange woods springing from the teeming earth] relate how heretofore,
> When ancient Pict began to whore,
> Deluded of his assignation
> (Jilting, it seems, was then in fashion),
> Poor pensive lover, in this place
> Would frig upon his mother's face[21]

Instead of a copy of Amphion's civilised Thebes, the reader is encouraged to see in the park a useful space for the satisfaction of sexual appetite. The masturbatory acts of the jilted Pict generate the rows of mandrakes that manage to transform themselves into the dense foliage that at present facilitates the promiscuous sexual mingling of bodies in the ironically consecrated spaces of the 'sin-sheltering grove'.[22] This perverse play on the topos of generation—accounting for the birth of the park with an account of onanism and incest—parodies the ropes of historical transmission initially constructed by the narrator's ironic deployment of the conventions of mythopoeic historiography.

As Jonathan Sawday has noted, the return of monarchy in 1660 was an 'unprecedented event in British history', an event that rendered problematic the notion of history itself and the tools enabling its representation.[23] In Sawday's terms, the end of the Protectorate and the restoration of the Stuart dynasty precipitated a representational crisis: 'Was History, in other words, to be considered as starting again, or was it still a continuum of ordered change?'[24] First, 'Restoration' itself cannot be understood without understanding how the very name incorporates the fact of a rupture, however that rupture is itself considered. That rupture is nothing other than the English Revolution or English Civil Wars of 1642–1649, which culminated with the legal trial and execution of Charles I, and the establishment of the English Commonwealth. The paradox of the name is that it cannot help but betray what it precisely refuses: the fact and consequences of the Interregnum. The 'Restoration', moreover, at once denominates an event and the period immediately following this

---

[21] Wilmot 1999, Lines 13–18.

[22] Wilmot 1999, Line 25.

[23] Sawday 1992, 171.

[24] Sawday 1992, 171.

event—but without providing any further predicates to specify what, beyond a dynasty, has been restored. Rather, the Restoration thereby and therefore clearly becomes the name of an unprecedented problem—how does one act in circumstances which have literally overthrown all established verities, and which essay to rupture with the rupture that immediately preceded them?[25] If a king can be tried and executed, and monarchy replaced by a commonwealth, then it is clear to everybody that new solutions have to be found: political, philosophical, and ideological. These solutions, moreover, will have to confront directly the massive evidence of political contingency. Such contingency opens an immediate double problem: that of legitimacy, and that of transmission. For if tradition has been proven to lack the necessary powers to ensure not only the handover of power, but of established power's continuing existence, then all bets are off. Charles II was required to battle for monarchy's legitimacy and for his lineage's transmission. Rochester's *Ramble* can be read in light of the representational crises created by the recent ruptures and attempted restorations of secure political transmission.

The sense of sacrilege conveyed through the narrator's deliberate manipulation of sacred rhetoric points toward the discrepancy between word and thing, calling attention to the manipulative nature of language and its ability to transform base appetite into sacred history. The poem's ironic handling of the act of consecration plays on the park's inherited religious provenance and its royal lineage, suggesting that the inherited honour the park acquires from its patron saint and perhaps from present royal authority serves as a convenient rhetorical veil concealing the dishonourable acts performed in the park's shady recesses. If rhetoric conceals and masks private appetite on the one hand, it also possesses the potential to unmask and reveal, however. The narrator disassociates the park from its honourable saint, but does so by way of offering its own 'consecrating' myth. The 'loved folds of Aretine' are substituted for the honourable St. James, with rhetorical convention functioning in the poem as both obfuscation and revelation of private motives and desires.

The sense of demystification is continued as the narrator begins a catalogue of persons who take advantage of the park's 'loved folds' to indulge in their promiscuous swiving. Parodying Waller's copious cataloguing of the park's attractions, Rochester's narrator squeezes great ladies and chambermaids, ragpickers and heiresses into single poetic lines, rewriting Waller's catalogue of pastoral pleasures as scenes of base carnality. The image of indiscriminate swivers seeking anonymity

---

[25] It is striking to us that 'Restoration' has so rarely been thought as a philosophical concept. In this regard, the remarks of Alain Badiou in a French context are at once illuminating and somewhat lacking insofar as the English elements are not considered: 'Since a restoration is never anything other than a moment in history that declares revolutions to be both abominable and impossible, it comes as no surprise that it adores number, which is above all the number of dollars or euros […]. Most importantly, every restoration is horrified by thought and loves only opinions; especially the dominant opinion, as summarized once and for all in François Guizot's imperative: "Enrich yourselves!" The real, as the obligatory correlate of thought, is considered by the ideologues of restorations—and not entirely without reason—as always liable to give rise to political iconoclasm, and hence Terror. A restoration is above all an assertion regarding the real; to wit, that it is always preferable to have no relation to it whatsoever' (Badiou 2007, 26).

amidst the foliage of the sacred walks works to destabilise social hierarchy, revealing the double-edged potential of linguistic and rhetorical convention to facilitate as well as undermine the production and transmission of social identities. The narrator's account of the promiscuous sexual liaisons enabled by the dark shelter of the park resembles the conventions of secret history, narratives that traditionally promised the revelation of base desires that lay under the cover of public personas. The narrative conventions of the secret history had an especially destabilising force in the mid-century political crisis, where they were deployed often in the political pamphleteering unleashed in the English civil wars, used specifically to challenge figures of political authority, with the secret ambitions and perversions of great men from Oliver Cromwell to Charles I displayed before a reading public now encouraged to discover perverse appetites lying beneath public masks of social and political authority.[26]

Immediately following the levelling catalogue of 'buggeries, rapes, and incests', the narrator quickly returns sacred rhetoric to the poem, re-describing the 'sin-sheltering grove' as a series of 'hallowed walks'. Introducing Corinna, the narrator resorts to conventional etiology once again, painting Corinna's beauty in broad Petrarchan strokes, with her presence in the park ascribed to her cold rejection of a despairing god's romantic suit. The poem thus seems to derive its impulse less from the desire to unmask the private secrets of public political figures than from a compulsive desire to stage repetitive scenes of veiling and unveiling.[27]

Corinna's subsequent spurning of the narrator echoes elements from the earlier 'myth' of the frigging Pict, whose fertile generation involved a similar experience of being cheated out of an assignation by a jilting lover. The association of 'jilting' with 'fashion' points to the poem's recurring concern with the problem of convention and the human appetites it is able to curb—a concern that generates the repeated rhetorical acts of sanctification and demystification, masking and unveiling. When read against Waller's vision of a prosperous community under the virtuous government of a disciplined, temperate monarch ruling by divine decree, Rochester's attention to the complex relationships between social convention and private indulgence, social intercourse and bestial appetite, arguably offers deeply unsettling questions on the nature of sociability, rejecting Waller's traditional vision of public virtue grounded in moral temperance in favour of a more unrelenting interrogation of the basis of community and the grounds of common consent to social convention and political authority. The poem's repeated staging of levelling and sacralising can be read as a discursive engagement with the problem of 'restoration'—the problematic return of a formerly rejected form of political authority and the contested terms for its 're-legitimation'.

---

[26] On the secret history as a product of the seismic political shifts enacted by the mid-century civil wars, see McKeon 2005, 469–505; on the effects of the tropes of revelation and discovery used in political discourse, see Achinstein 1994, 149–172.

[27] For an interesting discussion on Rochester's use of highly conventional generic categories to convey his destabilising, levelling narratives, see Sanchez 2005, 441–459.

In its emphasis on sexual appetite as the open secret of public institutions—the poem makes explicit its debts to the controversial Hobbesian idea of the social contract being predicated on the fundamentally asocial nature of man. Hobbes's own materialist reduction of human reason to the sum of one's appetites and passions was an attempt to solve the crisis in political obligation generated by the growth of competing accounts of political authority and subjection.[28] The poem's tenuous attempts to order an indiscriminate ramble with trite rhetorical formulas draw attention to the stabilising work performed by generic convention while revealing its arbitrary force. As Sharon Achinstein has noted, Hobbes's social theory posits both the urgent necessity for an absolute interpretive authority and the impossibility of associating truth with any form of linguistic practice.[29]

Rochester's reduction of the social exchange in St. James's Park to promiscuous swiving follows Hobbes's radical notion of man in a state of nature ruled solely by appetite. Hobbes's proposal for the restoration of political order involved acknowledging the asocial nature of man's natural drives and the fabricated and fictional underpinnings of human community. Consent to political authority, for Hobbes, was not the result of man's inherently social and political nature or the dictates of natural law, but rather, the result of a social contract that would have individuals acknowledge the base, bestial nature of man and the need to restrain and curb his natural liberty in order to prevent a war of all against all. Rather than reasserting the traditional myths of man as a social creature subjected to the dictates of a universal *ratio*, Hobbes's response to England's mid-century political crisis was to advance an idea of the social contract that presupposed the absence of sociability and community in the motives for political association. The anarchic nature of man was not only the source of political chaos, but of a new form of radical order.[30]

Rochester's libertine engagement with Hobbesian accounts of natural man, with its levelling of social hierarchies and the unmasking of convention, comprises part of an ongoing dialogue on the nature of authority and obligation in the Restoration. The libertine philosophy, cultivated in a courtly circle of wits, continued to build on Hobbes's overthrow of traditional moral philosophy and his elevation of fear as the source of the social and political contract, the initiator of government. Celebrating the Hobbesian levelling of men to a condition of total equality in the state of nature, Restoration libertinism saw social convention and its governing authorities as providing what Christopher Tilmouth describes as a 'controlled forum within which men could exercise their appetites (not least for power) even whilst maintaining a

---

[28] On the specifically political nature of Hobbes's treatment of the passions, see Strauss 1963; Tilmouth 2007, 257–313; Tuck 1993, 137–138.

[29] Achinstein 1994, 96–101.

[30] According to Tuck, 'Hobbes by 1651 [...] was a kind of utopian. *Leviathan* is not simply (and maybe not at all) an analysis of how political societies are founded and conduct themselves. It is also a vision of how a commonwealth can make us freer and more prosperous than ever before in human history, for there has never yet been a time (according to Hobbes) when the errors of the philosophers were fully purged from society, and men could live a life without false belief'. Tuck 1993, 137–138.

show of civility'.[31] Carolean court culture furthered a Hobbesian concept of liberty, a liberty shorn of its traditional links to the cultivation and free reign of Algernon Sidney's trio of 'law, justice [and] truth', a reductive form of liberty associated with the constraint of perpetually insatiable appetites.[32] Rather than seeing an autonomous community as a union of individual efforts to attain temperate self-government over private passions for the common good, the libertine ideal of freedom proposed a principle of restraint that understood the curbing of the passions as an instrumental necessity in order to procure a limited space for the indulgence and satisfaction of individual appetites—a licensed licentiousness—the source of the new social contract that for a republican like Sidney, could bind only 'villains'.

In 'A Ramble', Rochester furthers Hobbes's exploration of the relationship between appetite and convention. The poem's acts of unveiling reproduce the uncompromising frankness of Hobbes's materialist reduction of man to the sum of his roving appetites. The poem's guiding action of rambling, as the narrator moves from tavern to park, from inebriation to lust, embodies the perpetual motion of natural drives as his appetites take him from one site of corporeal indulgence to another. At the same time, however, the demystifying impulse of the poem's reduction of human action to indiscriminate rambling is conveyed through the heavily regulated form of the couplet and the highly conventional generic registers of mythic historiography and pastoral romance. Promiscuity and bestial pleasures are at the very heart and centre of social convention rather than its hidden secret.

While Rochester's poem works by revealing bestial appetites as the foundation for civic achievements, his poetry at the same time acknowledges the problem of establishing order and convention out of a bestial state of nature where man is a wolf to man. Having reduced the visitors of St. James's hallowed grounds to a swarm of promiscuous swivers, the narrator elevates one of its visitors, the jilting Corinna, to divine heights, describing her beauty in the idealising tropes from the pages of pastoral romance. Rather than facilitate the narrator's desired ramble, however, Corinna snubs him, preferring the company of men the narrator contemns for their foppishly excessive regard for the conventional postures of love and courtship.

Recent critical readings of *A Ramble* have attempted to make sense of the abrupt rage the narrator unleashes upon Corinna after he is snubbed, traditionally reading the narrator's sudden outburst as an anxiety-ridden response to the breakdown in social order, represented here by Corinna's inability to discriminate between the narrator's genteel status and that of the three mobile upstarts, whose company she prefers. Mona Narain has described Corinna as an abject figure embodying the threat of the anarchic dissolution of rank, a figure representing a new form of class mobility, who has to be contained and disciplined by the male, aristocratic narrator.[33] Reba Wilcoxon argues that the threat Corinna poses to the narrator's sense of order derives from her indiscriminate promiscuity, leading

---

[31] Tilmouth 2007, 259.
[32] Sidney 1698, 326. On the redefinition of liberty advanced by Hobbes and its divergence from classical notions of republican liberty, see Skinner 2008, 127–128.
[33] Narain 2005, 560–562.

the narrator to attempt an abusive mastery over her threateningly indiscriminate body.[34] We contend, however, that the very converse of this is true: Corinna functions less as the proverbial leaky vessel upsetting the narrator's sense of social stability, and more as a troublingly *discriminating* figure who, in fact, furthers the Hobbesian logic initially espoused by the narrator of the poem. This furthering is accomplished by her introducing a principle of discrimination into the promiscuous rambling that has defined the narrator and his verse. For what upsets the narrator is the opposition Corinna offers to his rhetoric of social levelling in her rejection of rambling, 'mere lust'. As the narrator rants:

> Such natural freedoms are but just:
> There's something generous in mere lust.
> But to turn damned absolute abandoned jade
> When neither head nor tail persuade…[35]

Corinna has turned from a rambling libertine, generous in her indiscriminate lust, into a calculating 'jade'—from promiscuous swiver into a canny whore in understanding. What provokes the narrator's anger as he downgrades Corinna from rover to prostitute is Corinna's preference for the company of a trio of males who are described by the narrator in terms of *their* desires and potential for social advancement. The narrator, in a tone of nostalgic lament, reminisces on his 'dissolution' in Corinna's breast, as he recounts the pleasures he takes in her 'lewd cunt' 'drenched with the seed of half the town'.[36] The dissolution the narrator prizes is the pleasure taken in mixing indiscriminately his 'dram of sperm' with the 'slime' drawn from half the town, from porters to footmen. Interestingly, his tirade against Corinna's abandonment of promiscuous pleasure soon moves into a curse, which anticipates her future marriage. Rather than being a threat to established rank and hierarchy, Corinna instead functions for the narrator, we argue, as a figure of contract and convention. She embodies the principles structuring the necessary and voluntary exit out of a state of rambling nature and into a contract of mutual obligation enabled by the discipline one exerts over one's appetites—the sacrifice of pure pleasure for the long-term considerations of a socially determined interest.

If Hobbes's theory of man reduces the human will to the sum of roving appetites, his theory of society involves the transformation of promiscuous pleasure-seeking into a form of rational calculation to secure long-term satisfaction. Having transformed Waller's pinnacle of civil achievement into an obscure shade for the satisfaction of private appetites, the narrator finds in Corinna the principle of a specifically Hobbesian restraint, the bridling of appetites that enable a form of order, signified in the promise of wedlock, opposed to the narrator's promiscuous ramble. Interestingly, the tirade against Corinna's discrimination is followed by an anticipation of her future marital problems. If perpetual promiscuity necessitates the social contract, it also hollows it out of any substance beyond the consideration and calculation of

---

[34] Wilcoxon 1976, 277.
[35] Wilmot 1999, Lines 97–100.
[36] Wilmot 1999, Line 113, 114.

one's pleasures. If the nature of man is reduced to indiscriminate motion necessitating the formation of a commonwealth, Rochester questions the permanence of such a formation, given that the only source of social obligation stems from the private consideration of individual appetites. It is the calculative move on the part of Corinna to leave off indiscriminate rambling that contributes a great deal to the narrator's misogynist tirade.

The three men Corinna privileges over the narrator are described in terms of their shameless pursuit of interests, in contrast to the narrator's liberal and cavalier indulgence of his sexual appetite. The three men are described in non-sexualised terms, unlike the other park swivers, and are instead associated with the theatrical. The first man is all about 'abortive imitation' and loving by 'rote'. The theatrical manipulation of the passions introduced into the park by the 'knights of the elbow' disrupts the dissolution of rank and status the narrator achieves in his earlier lines cataloguing the park's primary attractions. The second 'knight' is described as one who courts women using borrowed lines from stage plays, again suggesting the move from uninhibited rambling to a theatrical management of the appetites. The third 'knight' is an heir to a considerable estate, seeking to acquire the reputation and social capital of a libertine wit through association with the other two poseurs in the trio. While the narrator's disgust for the upstart trio can be read as an attempt to maintain a distinction between a genuine aristocratic magnanimity and an emerging set of bourgeois mercenary interests, the narrator's own demonstrated drive to reduce the artifice of form to roving, will-less appetite complicates the kind of classed distinction between a noble generosity and the self-denying management of desires for the sake of petty profit.[37] The sense of aristocratic privilege posited by some critics in their readings of Rochester's condemnation of the trio of 'knights' strikes a discordant note in a narrative that seems bent on destroying the possibility and desirability of a classed position. The anguish caused by Corinna's discriminating jilt gestures, this essay argues, towards the difficulties inherent in Hobbes's redefinitions of the social and political contracts, and their application to a mode of practical ethics. If, as Quentin Skinner has argued, Hobbes was positing an alternate vision of the good life founded upon a reconsideration of the principles of liberty and restraint, Rochester, in his engagement with Hobbesian equality, is gesturing to the limits of the Hobbesian social contract and its capacity to generate a substantive mode of ethics.

3.3 Conclusion

We have tried to show that Rochester's life and work constitute an unprecedented, intimate, and subtle demolition of the political arguments of Hobbes. This demolition takes a surprising form: first, of near-total identification of Rochester with the

---

[37] On aristocratic magnanimity as a defensive move against incursions into traditional class privilege in the later seventeenth-century, see Tilmouth 2007, 315–370; Scodel 2002, 247–251.

Hobbesian program; second, of an unparalleled attentiveness to the surprising, unintended consequences of this program; third, of the presentation of this relation in the mode of a new kind of poetic satire. As we have argued, Rochester's work also necessarily dissimulates its own true relationship to Hobbes; this is a positive feature of his work rather than a failure. The key aspects of Rochester's experimental ethical critique primarily concern the consequences for sociability: if, as Hobbes argues, only the sovereign can ensure that the many can come together as one to form any viable political community, Rochester shows how this structure entails the nightly return of a radically egalitarian eroticism. Moreover, Rochester protests that, from within this eroticism itself, a variety of new, post-aristocratic modes of social evaluation emerge as a matter of *physical* course, whereby, through the simple motion and encounter of sovereignly pliable bodies, derisory-yet-unexcludable attempts to exercise ungrounded discriminations as a form of control arise in bodies and strive to elaborate themselves. It is this 'excess' that Rochester recognises that the Hobbesian sovereign *cannot not* license. Rochester is thus irreducibly ambivalent about Hobbes: an absolute adherent, yet, in being so, he realises that such sovereign power breeds the phantoms of a new darkling world.

Rochester can therefore be considered an unapparent precursor of sensibility in at least two senses. The first is in his negative political demonstrations: Rochester's work exposes the necessity of a form of government that can capture precisely what Hobbes's not only cannot, but induces as if automatically—a risible erotic hierarchising that is at once utterly demotic yet lays claim to a privilege it precisely destroys (this is the 'theatre of criticism' that the *Ramble* poetically stages). The second is also negative, but hinges on the order of explanation: it demands that a new *account* be given of the functionings of erotic judgement. If the definitive solution to the failures of Hobbes is given by Locke, and, after him, by thinkers such as Shaftesbury or, in France, the *philosophes*, the key to their very different proposals quite sensibly comes to be sense itself: common sense or good sense, sense and sensibility. These terms thereafter come to be linked, in all sorts of semantic and social registers, with others, such as 'politeness'.[38] For what these terms do—or at least one shared thing that they attempt to do—is bind psychophysiology to sociability in a mode that accounts for the necessity of human variation, yet simultaneously legitimates certain forms of moralised governmental unification.[39] Precisely as Anne Vila says in *Enlightenment and Pathology*, 'the various meanings attached to sensibility tended to be mutually permeable because eighteenth-century authors used the word as a bridging concept—a means of establishing causal connections between the physical and the moral realms'.[40] But it is precisely to the necessity of such a mediating term that Rochester points, as he shuts down the strongest then-existing account of the transition from nature to sovereignty in materially affirming all its implications. In doing so, he became—and thereafter remained—a great negative example, fêted by Voltaire himself in

---

[38] See Klein 1994.
[39] See McKeon 2005.
[40] Vila 1998, 2.

the famous *Lettre XXI* of *Lettres philosophiques*, which can serve as a fitting conclusion to this essay: 'The Earl of Rochester's name is universally known. Mr de Saint-Evremond has made very frequent mention of him, but then he has represented this nobleman in no other light than as the man of pleasure, as one who was the idol of the fair; but with regard to myself, I would willingly describe in him the man of genius, the great poet'.[41] Rochester's genius is certainly evident in his vitiation of the master he follows to the very end, and in contributing to making possible the sensibility he would have abominated.

References

Achinstein, Sharon. 1994. *Milton and the revolutionary reader*. Princeton: Princeton University Press.
Badiou, Alan. 2007. *The Century*. Trans. Alberto Toscano. Cambridge: Polity.
Bartlett, A.J. 2011. *Badiou and Plato: An education by truths*. Edinburgh: Edinburgh University Press.
Berman, Ronald. 1964. Rochester and the defeat of the senses. *The Kenyon Review* 26(2): 354–368.
Bloom, Harold. 1973. *The anxiety of influence: A theory of poetry*. Oxford: Oxford University Press.
Bloom, Harold. 1975a. *Kabbalah and criticism*. New York: Continuum.
Bloom, Harold. 1975b. *A map of misreading*. Oxford: Oxford University Press.
Bloom, Harold. 1982. *Agon: Towards a theory of revisionism*. Oxford: Oxford University Press.
Chernaik, Warren. 1995. *Sexual freedom in restoration literature*. Cambridge: Cambridge University Press.
Combe, Kirk. 1998. *A martyr for sin: Rochester's critique of polity, sexuality, and society*. Newark: University of Delaware Press.
Courthorpe, William. 1903. *A history of English poetry*. London: Macmillan.
Cousins, A. 1984. The context, design and argument of Rochester's *a Satyr against reason and mankind*. *SEL: Studies in English Literature* 24(3): 429–439.
Farley-Hills, David (ed.). 1972. *Rochester: The critical heritage*. London: Routledge.
Fujimura, Thomas. 1958. The originality of Rochester's *Satyr against mankind*. *Studies in Philology* 55(4): 576–590.
Greene, Thomas M. 1982. *The light in troy: Imitation and discovery in renaissance poetry*. New Haven: Yale University Press.
Johnson, Ronald. 1975. Rhetoric and drama in Rochester's *a Satyr against reason and mankind*. *SEL: Studies in English Literature* 15(3): 365–373.
Kahn, Victoria. 2004. *Wayward contracts: The crisis of political obligation in England, 1640–1674*. Princeton: Princeton University Press.
Klein, Lawrence. 1994. *Shaftesbury and the culture of politeness: Moral discourse and cultural politics in early eighteenth-century England*. Cambridge: Cambridge University Press.
Knight, Charles. 1970. The paradox of reason: Argument in Rochester's *Satyr against mankind*. *Modern Language Review* 65(2): 254–260.
Lacan, Jacques. 1966. *Ecrits*. Paris: Seuil.
McKeon, Michael. 2005. *The secret history of domesticity: Public, private, and the division of knowledge*. Baltimore: Johns Hopkins University Press.

---

[41] Farley-Hills 1972, 194.

Moore, John F. 1943. The originality of Rochester's *Satyr against mankind. PMLA* 52(2): 393–401.
Narain, Mona. 2005. Libertine spaces and the female body in the poetry of Rochester and Ned Ward. *ELH* 72(3): 553–576.
Parkin, Jon. 2007. *Taming the leviathan: The reception of the political and religious ideas of Thomas Hobbes in England 1640–1700*. Cambridge: Cambridge University Press.
Parsons, Robert. 1681. *A sermon preached at the funeral of the right honourable John Earl of Rochester*. Dublin: Printed by Benjamin Took and John Crook, and are to be sold by Mary Crook.
Robinson, K. 1973. Rochester and Hobbes and the irony of *a Satyr against reason and mankind. The Yearbook of English Studies* 3: 108–119.
Russell, Ford. 1986. Satiric perspective in Rochester's *a Satyr against reason and mankind. Papers on Language and Literature* 22(3): 245–253.
Sanchez, Melissa. 2005. Libertinism and romance in Rochester's poetry. *Eighteenth Century Studies* 38(3): 441–459.
Sawday, Jonathan. 1992. Re-writing a revolution: History, symbol and text in the restoration. *The Seventeenth Century* 7(2): 171–199.
Scodel, Joshua. 2002. *Excess and the mean in early modern English literature*. Princeton: Princeton University Press.
Sharpe, Kevin. 1987. *Criticism and compliment: The politics of literature in the England of Charles I*. Cambridge: Cambridge University Press.
Sidney, Algernon. 1698. *Discourses concerning government*. London: John Toland.
Skinner, Quentin. 2008. *Hobbes and republican liberty*. Cambridge: Cambridge University Press.
Strauss, Leo. 1963. *The political history of Hobbes, its basis and its genesis*. Chicago: University of Chicago Press.
Tilmouth, Christopher. 2007. *Passion's triumph over reason: A history of the moral imagination from Spenser to Rochester*. Oxford: Oxford University Press.
Treglown, Jeremy. 1973. The satirical inversion of some English sources in Rochester's poetry. *The Review of English Studies* 24(93): 42–48.
Tuck, Richard. 1993. The civil religion of Thomas Hobbes. In *Political discourses in early modern Britain*, ed. Nicholas Phillipson and Quentin Skinner, 120–138. Cambridge: Cambridge University Press.
Turner, James G. 2002. *Libertines and radicals in early modern London*. Cambridge: Cambridge University Press.
Vila, Anne. 1998. *Enlightenment and pathology: Sensibility in the literature and the medicine of eighteenth-century France*. Baltimore: Johns Hopkins University Press.
Waller, Edmund. 2001. On St. James's Park, as lately improved by his majesty. In *The broadview anthology of seventeenth-century verse and prose*, ed. Joseph Black, Holly Faith Nelson, and Alan Rudrum, 500–503. Peterborough: Broadview Press.
Webster, Jeremy. 2005. *Performing libertinism in Charles II's court*. Basingstoke: Palgrave Macmillan.
Wilcoxon, Reba. 1976. The rhetoric of sex in Rochester's burlesque. *Papers on Language and Literature* 12: 273–284.
Wilmot, John, and The Earl of Rochester. 1999. A Ramble in St. James's Park. In *The works of John Wilmot, the Earl of Rochester*, ed. Love Harold, 76–80. Oxford: Oxford University Press.

Chapter 4
Emotional Sensations and the Moral Imagination in Malebranche

Jordan Taylor

Abstract This paper explores the details of Malebranche's philosophy of mind, paying particular attention to the mind-body relationship and the roles of the imagination and the passions. I demonstrate that Malebranche has available an alternative to his deontological ethical system: the alternative I expose is based around his account of the embodied aspects of the mind and the sensations experienced in perception. I briefly argue that Hume, a philosopher already indebted to Malebranche for much inspiration, read Malebranche in the positive way that I here describe him. Malebranche should therefore be acknowledged as a serious influence on Enlightenment philosophy of sensibility.

> Briefly, man's life consists only in the circulation of the blood, and in another circulation of his thoughts and desires. And it seems we can hardly use our time better than in seeking the causes of these changes that happen to us, thereby learning to know ourselves.[1]

In one of his *Philosophical Letters*, first published in 1731, Voltaire paints a rather damning picture of Malebranche:

> M. Malebranche, of the Oratory, in his sublime hallucinations, not only allowed the existence of innate ideas but was certain that all we perceive is in God and that God, so to speak, is our soul.[2]

Voltaire's interpretation of Malebranche is simply wrong. Firstly, Malebranche quite explicitly *rejected* the existence of innate ideas—it was one of his key criticisms of the Cartesian account of knowledge. To this end, Malebranche devoted

---

[1] Malebranche 1678/1997, 90.
[2] Voltaire 1733/2003, 52.

J. Taylor (✉)
Department of Philosophy, University of Pennsylvania,
433 Cohen Hall, Philadelphia, PA 19104, USA

Department of Cognitive Science, Macquarie University, North Ryde, NSW, Australia
e-mail: jordt@sas.upenn.edu

a chapter of Book III of his magnum opus, *The Search after Truth*.³ Secondly and more subtly, Malebranche does not hold that we perceive *all* things in God. Granted, all the *truths* of the external world we gain through our pure perceptions of eternal ideas are found in God. But in his single-sentence dismissal of Malebranche, Voltaire entirely ignores the internal world of the embodied mind: a world of sensations, passions, and, I will demonstrate, sympathy or compassion.

Malebranche's epistemological system splits our means of experiencing the world into two distinct classes: pure perceptions of ideas⁴ and sensations or sentiments. These latter terms, *sensations* and *sentimens*, refer to the same type of thing throughout Malebranche's works, and they are typically translated and treated in Anglophone literature as 'sensations'—I follow suit. The class of sensations can be further divided into two subclasses: perceptual sensations such as colours and flavours, pleasures and pains, and emotional sensations such as joy and sadness. In this paper, I demonstrate the ways in which Malebranche distinguishes these two types of sensations, and why such distinctions are important to his system. In doing so, I emphasise a point about Malebranche's mind-body dualism that is often ignored by those who seek to characterise negatively positions such as his: sensations and passions are demonstrative of an embodied mind.⁵

I explore Malebranche's theory of perception and the passions, and towards the end of the paper I note some of the theory's influences on David Hume's works. My aim is to demonstrate one of the ways in which Hume utilised Malebranche's theory of the passions and the mind's natural inclination towards compassion, arguing that, despite notable incompatibilities in their ethical commitments, the two philosophers have more in common than is often acknowledged. Key to understanding this commonality is Malebranche's account of mind-body interaction; I therefore explain at some length his treatment of sensory perception, the imagination, and the passions. In the first section, I describe what Malebranche calls pure perceptions. These are acts properly attributed to the disembodied or meditative mind whose purpose is to provide the mind with eternal truths about the intelligible world. Since Malebranche is primarily concerned with attaining truths, his emphasis is on pure perceptions of ideas throughout the *Search*. The fact that it is through pure perception that we discover truths sees Malebranche write of sensory perception rather negatively: since perceptual sensations do not afford us access to eternal and necessary truths, they

---

³ See Malebranche 1678/1997, 226–227. For commentary, see Schmaltz 1996, 96–99; Jolley 1988.

⁴ I do not offer an interpretation of Malebranche's theory of ideas here. It should suffice to say that, for Malebranche, ideas are intelligible representations of objects perceived externally by the mind; in many respects, they are similar to Plato's Forms. By virtue of their being external and abstract, they are what give rise to our purely objective knowledge; they differ from sensations not only ontologically, but epistemologically—ideas are not *thoughts*; rather they are *thought of*.

⁵ I use the term 'embodied' in a qualified sense throughout this paper. In Malebranche's system, a mind and a body are metaphysically distinct, since they are composed of different substances which do not causally interact. But the mind and body are intimately connected, both functionally and phenomenologically: actions of the mind and body correspond to one another, and the movements of the body give rise to sensations in the mind. (This is explained in more detail below.) It is in this sense that the term 'embodied mind' is employed.

are not very helpful in our intellectual investigations into the world. I then compare this pure perception with Malebranchean sensory perception, explaining the practical, scientific[6] function of perceptual sensations before linking them with their physical counterparts in the body's sensory organs and brain: depending on circumstances, perceptual sensations arise due to the senses or the imagination. Next, I turn to the passions and the emotional sensations they provide. In the penultimate section, I show some of the implications of the ways in which the imagination and the passions influence one another. These implications lie dormant in Malebranche's work, but they are demonstrative of some of the positive contributions to life on the part of the passions, the imagination, and sensory perception, all of which can be considered as activities of the embodied mind. In the final section, I offer a kind of case study of Malebranche's positive influence on Enlightenment notions of sensibility by demonstrating that Hume noticed these implications in his own reading of Malebranche and adapted them to his own purposes. Readers should note, however, that the discussion of Hume is brief and suggestive, rather than detailed; this is not a paper about Hume. Rather, by explicating his theory of the passions, I hope to show that Malebranche deserves mention amongst the great influencers of the Enlightenment era, not necessarily as a target or deluded theologian, but as a thinker whose theory warrants positive and serious reading.

4.1 Pure Perceptions and the Disembodied Mind

In Malebranche's system, the mind or soul, an immaterial and unextended entity, is a very malleable creature. It is capable of changing in an indefinite (perhaps infinite) number of ways depending on what is acting upon it. Different stimuli—different ideas in God or, less directly, objects in the world[7]—trigger or correspond to different modifications of the mind. These modifications come in two forms: pure perceptions (*pures perceptions*) and sensations (*sensations* or *sentimens*). The former are concerned with truth; the latter, with the body.

Malebranche claims that it is because of the body that we fall into error in our understandings, as our attention is pulled away from the eternal truths revealed in intelligible, pure perceptions of ideas. The mind's union with its physical vessel renders it the slave of the body.[8] Malebranche's pessimism in the *Search* regarding the corruptibility of the mind by the body is motivated by the objective of the work: since truths are reached by means of pure perceptions, the seeker of truth—the intellectual mind—should 'be awakened from its somnolence and make an effort to free

---

[6] As we will see below, for Malebranche, the sciences do not yield truths of the same kind as do metaphysics or theology.
[7] Note that, for Malebranche, the body is an object in the world in the same way as are rocks and trees; see the discussion on passions below.
[8] Malebranche 1678/1997, xxxv.

itself' from the burdens of the body.[9] In other words, it is when the mind is disembodied that it is able to perceive truths.

It is important to note that, for Malebranche, all thought is some type of perception[10]; the mind's modifications being different ways of perceiving.[11] The 'understanding' (*l'entendement*) is the faculty of mind that receives all of its different modifications, where, by a 'faculty' of the mind, Malebranche simply means a capacity.[12] With respect to thought, ideas are not found *within* the mind: they are rather *thought of* or *perceived by* the mind, since to 'see nothing is not to see; to think of nothing is not to think'.[13] Ideas are perceived within God's pure intellect, to which our minds are intimately connected. (On this point, one must concede, Voltaire did get Malebranche right.)

In pure perception, we are able to perceive an eternal idea clearly and intelligibly. These pure perceptions do not make an impression on the mind, nor do they sensibly modify it.[14] Yet without *sensing* our pure perceptions, we are still aware of them through what Malebranche calls inner sentiment (*sentiment intérieur*) or consciousness (*conscience*). The different modifications of the mind

> cannot be in the soul without the soul being aware of them through the inner sensation it has of itself—[modifications] such as its sensations, imaginings, pure intellections, or simply conceptions, as well as its passions and natural inclinations.[15]

In effect, pure perceptions are means of perceiving ideas which render those ideas intelligible. It is through a pure perception of a triangle that we are able to deduce its mathematical properties; all 'spiritual things, universals, common notions, the ideas of perfection and of an infinitely perfect being, and generally all its thoughts when it knows them through self-reflection' are perceived by means of pure perceptions.[16] It is through pure perception, then, that we are able to perceive abstract ideas, as well as relations between ideas (judgements) and relations between those relations (inferences).[17] In Malebranchean epistemology, judgements and inferences, just like ideas themselves, are not *made* so much as *perceived*: they are themselves pure perceptions.[18] Judging is perhaps best understood as perceiving two ideas through the same pure perception or modification of mind, thereby yielding a perceived relation between those two ideas. To judge that 'two

---

[9] Malebranche 1678/1997, xxxix.

[10] Simmons 2009, 105–129.

[11] Malebranche 1678/1997, 2.

[12] Malebranche 1678/1997, 3.

[13] Malebranche 1678/1997, 320.

[14] Malebranche 1678/1997, 2. Our minds are, however, modified by way of 'pure intellections'. A discussion of what this entails will take us too far from our present topic, but for a detailed and careful analysis see Jolley 1994.

[15] Malebranche 1678/1997, 218.

[16] Malebranche 1678/1997, 16.

[17] Malebranche 1678/1997, 7–11.

[18] Malebranche 1678/1997, 7.

times two is equal to four' is to notice a relation of equality between the idea of 'two times two' and the idea of 'four'. Inferring is the act of perceiving relations between two judgements: as we judge that 'six' is greater than 'four', we infer that it is also greater than 'two times two' (by virtue of the previously noticed relation between 'two times two' and 'four'). Such relations between ideas define Malebranche's notion of truths:

> Now, truths are but relations of equality or inequality between these intelligible beings (since it is true that twice two is four or that twice two is not five only because there is a relation of equality between twice two and four, and one of inequality between twice two and five).[19]

It is through pure perceptions of ideas, then, that we gain any truths about the intelligible world. Interestingly, a truth is not found *in* an idea, but rather in the mind's own pure perception of two ideas. Malebranche further distinguishes between three kinds of truths: truths as relations between ideas (such truths are metaphysically necessary), as relations between ideas and corresponding things in the world, and finally, as relations between different things in the world.[20] What pure perception offers is a means of intelligibly perceiving or thinking of things—a way of making sense of the world outside the human mind.

Due to the fact that these pure perceptions attend only to the ideas present in the intellectual realm, independently of anything material,[21] they are in no way dependent upon the body. When the mind knows objects by pure perception alone, 'without forming corporeal images of them in the brain to represent them',[22] it perceives them as purely abstract and universal. But because abstract thoughts neither rely upon nor excite the body, the mind views them as remote and struggles to apply itself to them.[23]

Seekers of eternal and necessary truths, then, are burdened by their bodies. This pessimistic perspective on the human state resonates throughout Malebranche's work. Yet it would be a mistake to say on Malebranche's behalf that its union with a body is entirely detrimental to the human mind. Abstract truths often do not reflect the here-and-now situations in which we (in our bodies) find ourselves, and to which we must react. In fact it is not, strictly speaking, our bodies which lead us to error and away from truth; it is rather the will that leads the mind astray, conceding to sensible pleasures (and maintaining a cautious vigil against pains) before seeking epistemic clarity. Indeed it is not only in its union with the intellectual realm of God, but also its union with the material body, that a mind can be considered a complete

---

[19] Malebranche 1678/1997, 617–618.
[20] Malebranche 1678/1997, 433. It is to this first species of truth—the eternal and necessary truths of the intellectual realm—that Malebranche's use of 'truth' typically refers in the *Search*. Throughout this paper, I follow Malebranche's use of the term, though exceptions will be noted.
[21] Malebranche 1678/1997, 16–17.
[22] Malebranche 1678/1997, 198. The relationship between corporeal images and perceptions is explained below.
[23] Malebranche 1678/1997, 59, cf. 213.

person: Malebranche refers to the mind and body as 'the two parts of man'.[24] The next section is therefore an exploration of the practical, world-centric side of Malebranchean epistemology and science: perceptual sensations and sensory perception, the activities of the embodied mind.

4.2 Perceptual Sensations and the Embodied Mind

Competing with pure perceptions for our attention are our perceptual sensations, which 'make a more or less vivid impression' on the mind.[25] Examples include colours, flavours, heat/coldness, hardness, and pains.[26] They can be further distinguished as affective or non-affective sensations[27]: affective sensations such as heat and pains draw our attention directly to the body, while non-affective sensations such as colours are sensed as if in external objects so that those objects can be distinguished from one another.[28] Both types of perceptual sensation act much like alarm bells that ring in the presence of objects (or changes in the body) in order to draw the attention of the mind 'to preservation of its machine'.[29] Such perceptual sensations are bestowed upon us so that we can maintain the welfare of our bodies without having to draw too much of our attention to them and away from our pure perceptions of eternal ideas—at least that is their original, pre-lapsarian function. In Malebranche's account of the human being, the 'goods of the body do not deserve the attention of a mind' whose priority should always be to seek out truth. Sensations therefore provide indications of the presence of goodness or badness with respect to the body, in relation to the object impacting upon it:

> The mind, then, must recognize this sort of good without examination, and by the quick and indubitable proof of sensation. Stones do not provide nourishment; the proof of this is convincing, and taste alone produces universal agreement.[30]

In the pristine and peaceful Garden of Eden such perceptual sensations would be entirely reliable. Unfortunately, in our post-lapsarian state, we find ourselves in a hostile and volatile world, our bodies in constant danger from snakes, swords, and stubbed toes. Our attention is drawn more and more to the states of our bodies and we strive to attain good and avoid evil, with which we associate sensations of

---

[24] Malebranche 1678/1997, 52.

[25] Malebranche 1678/1997, 2.

[26] A pain, here, is taken to be that which comes with a physical wound or a headache. The sort of 'pain' that accompanies or constitutes emotional sensations is considered below.

[27] 'Affect' implies sensible pleasure or pain. Sensations accompanying wounds to one's body, headaches, or orgasms would all be considered affective sensations.

[28] Malebranche 1678/1997, 55. Malebranche does not dwell on this distinction as he realises that degrees of affect can vary across occasions; that is, it is not simply the case that some sensations are affective while others are not. See Malebranche 1678/1997, 57–58.

[29] Malebranche 1678/1997, 200.

[30] Malebranche 1678/1997, 21.

pleasure and pain respectively.³¹ In this context 'good' means good for the body: in fact, Malebranche maintains that what is good for the body is more often than not detrimental to the mind.³² As he reminds us, we do not experience sensations in order to perceive truths, but only so that we can preserve our bodies.³³ To make matters worse, this pull towards worldly pleasures leads us falsely to associate perceptual sensations (especially less affective sensations such as colours and tastes) with the perceived objects' causal relationships with the body. In other words, we fall into error when we judge worldly objects according to the sensations they evoke in the mind:

> When, for example, we see light, it is quite certain that we see light; when we feel heat, we are not mistaken in believing that we feel heat, whether before or after the fall. But we are mistaken in judging that the heat we feel is outside the soul that feels it.³⁴

The problem with judging that our perceptual sensations are qualities of perceived objects rather than modifications of the soul is that we begin to look for truths in those sensible qualities. We fall into error when we believe our sensations provide us with some truthful information about the nature of ideas. As Steven Nadler explains, our perceptual sensations, taken as the sensory qualities of colour, heat, and the like, 'possess no representational content, and contain no element of truth regarding the external world'.³⁵ That is, they cannot tell us anything about the nature of ideas—what real qualities they have—and as such, prove quite useless in the search after truths. As we saw in the previous section, truths (relations between ideas) are perceived through pure perception, an undertaking of the mind insofar as it can disembody itself, so to speak. It seems that since perceptual sensations provide no truths, they are not helpful to metaphysical enquiry. Malebranche offers a simple example of a perceptual sensation's potential to mislead:

> different objects can cause the same sensation of color; plaster, bread, sugar, salt, and so on, have the same sensation of color; nevertheless, their whiteness is different if one judges it other than through the senses. Thus, when one says that flour is white, one says nothing distinct.³⁶

A quality or property that triggers a perception of whiteness is something common across these otherwise unique materials, yet it is incorrect to say that such a relation is a truth. The fact that two materials evoke sensations of whiteness in no

---

³¹ Malebranche 1678/1997, 21. This correlation is also explained in Malebranche's quite questionable advice on raising children: he recommends against rewarding children with sensible pleasures as this will corrupt their motivations to learn and behave properly, steering attention towards bodily pleasures rather than reason. On the other hand, sensible punishments are justified in cases when children cannot be convinced through their own reason, as pain will impede children's enjoyment of vice and prevent the mind from being enslaved by the body. See Malebranche 1678/1997, 127–129.
³² See Malebranche 1678/1997, 62–63.
³³ Malebranche 1678/1997, 24.
³⁴ Malebranche 1678/1997, 23.
³⁵ Nadler 1992, 23.
³⁶ Malebranche 1678/1997, 442.

way explains or demonstrates a nature necessarily common to them both. This commonality 'is obscure because the same sensation of whiteness can be linked to objects with very different internal configurations'[37]: the likeness is merely contingent, even arbitrary. At best, we could qualify the common quality of whiteness as belonging to the third species of 'truth'. The truths perceived between eternal ideas, by contrast, are immutable and necessary (*immuables et nécessaires*) as are the ideas themselves.[38]

On the other hand, what sensations do provide are 'natural judgements' (*jugements naturels*) which 'are quite correct, if they are considered in relation to the preservation of the body', even if they are 'quite bizarre and far removed from the truth'.[39] Despite the fact that perceptual sensations do not represent real qualities of the world, they prompt us to react immediately, and typically appropriately, to the objects we encounter.[40]

Alison Simmons has recently argued that many Malebranche scholars have misread his position on sensations. Since Malebranche disallows sensations any representational content, they claim, he must likewise deny that sensations have any intentionality.[41] But as Simmons explains, a Malebranchean sensation, by virtue of being a way of thinking (as defined above), is certainly *about* or *of* something; it does more than simply add 'a bit of phenomenological panache' to an otherwise pure perception.[42] I agree with Simmons that Malebranchean sensations are non-representing yet intentional modes of the mind: they are ways of perceiving ideas, as are pure perceptions. However, I want to demonstrate the similarities between the intentionality of pure perceptions and that of perceptual sensations in a different way to Simmons. The explanation I offer revolves around the relationship between the body and the mind. My main claim is that perceptual sensations allow for perceptions of relations which in some ways resemble eternal truths, but are ultimately contingent, rather than immutable and necessary. This contingency is due to the fact that such a relation is not between two ideas, but between an idea and the body, throughout which the sensitive mind is embedded. Recall from above the second kind of 'truth' that Malebranche identifies: truth as a relation between idea and thing. This kind of truth, lacking the metaphysical necessity possessed by an eternal truth *qua* relation between ideas, is of greater interest to the natural scientist than it is to the metaphysician. (The same is true of the third kind of truth, truths as relations between things in the material world.) Thus, the sensations that afford us such *scientific* truths are useful to the embodied mind insofar as it interacts with the material world. To better grasp this claim, an explanation in Malebranche's terms of the psycho-physiology of sensory perception and imagination may prove helpful. This is offered in the two following sections.

---

[37] Schmaltz 1996, 58.
[38] Malebranche 1678/1997, 618.
[39] Malebranche 1678/1997, 60.
[40] Here Malebranche is echoing Descartes's position in his *Sixth Meditation*.
[41] Simmons 2009, 105.
[42] Simmons 2009, 110.

4.3 The Psycho-Physiology of Perception

Like Descartes, Malebranche is concerned with accounting for the physiological processes of the body as fundamental aspects of perception. However, armed with occasionalism on the one hand, and on the other, the correct observation that the biology and physics of his time is inadequate to develop an accurate and complex neurophysiology,[43] Malebranche is less concerned with demonstrating the exact psycho-physiological pathway from object in the world to perception by the mind, than with the functional relations between each step in the overall process of perception. His occasionalism calls for a rejection of metaphysical causal forces between the two substances of which we are comprised (extension and mind, or matter and thought).[44] So although Malebranche's physiology is incomplete, it can be read with an air of flexibility, a certain neurobiological agnosticism.

This flexibility, however, should not be taken as full liberty of explanation. Malebranche holds that there is an intimate, important connection between the composition of the body and the sensations of the soul. The matter constituting the body

> has to be flesh, brain, nerves, and the rest of a man's body so that the soul may be joined to it. The same is true of our soul: it must have sensations of heat, cold, color, light, sounds, odors, tastes, and several other modifications in order to remain joined to its body.[45]

This is backed up in the earlier chapters of Book I of the *Search*, where Malebranche paints a picture of the two substances as resembling one another in their modifications and capabilities. While he clearly states that it should only be taken figuratively, Malebranche relies heavily on a functional comparison between the different properties of each substance:

> Matter or extension contains two properties or faculties. The first faculty is that of receiving different figures, the second, the capacity for being moved. The mind of man likewise contains two faculties; the first, which is the *understanding*, is that of receiving various *ideas*, that is, of perceiving various things; the second, which is the *will*, is that of receiving *inclinations*, or of willing different things.[46]

It is thus understandable that there would be a close relation between the body's sensory system (composed of the sensory organs, the brain, the nerves linking them, and the animal spirits[47] running throughout the nerves and brain) and the faculty of

---

[43] Malebranche 1678/1997, 49–50; see also Sutton 1998b, 107.

[44] While this claim is straightforward enough for present purposes, debates continue over exactly how we should interpret Malebranche's doctrine of occasionalism. Nadler provides a good explanation of Malebranchean occasionalism in his article, 'Occasionalism and General Will in Malebranche', and offers a brief review of competing interpretations in his postscript to that article. Both pieces can be found in Nadler 2011.

[45] Malebranche 1678/1997, 200.

[46] Malebranche 1678/1997, 2.

[47] These are 'merely the most refined and agitated parts of the blood' which 'are conducted, with the rest of the blood, through the arteries to the brain', where 'they are separated from it by some parts intended for that purpose' (Malebranche 1678/1997, 91).

understanding. Thus, while the mind and body are separate entities on an ontological level, on a functional level they are interdependent and (almost) unified: what goes on in the body affects what is perceived by or in the mind, and the volitions of the mind determine many of the motions of the body. (I say *almost* unified because, of course, the body often moves of its own accord, and it is not possible for there to be any subconscious regulatory systems on the part of the mind which could account for these movements. Indeed, Malebranche does subscribe to the view of the body as sophisticated machine, capable of self-movement in a physically predetermined or dispositional sense.[48])

What happens in the process of (visual) perception can be described in the following way. First, the body encounters an object in the physical world—let's say an apple. The apple transmits rays of light in all directions, and these rays of light vibrate in such a way that they produce pressures in the various parts of the eyes.[49] Animal spirits agitated by these pressures then flow through the nerves of the eyes to the brain. If the object is novel, the animal spirits etch traces representative of, but not resembling, the object impacting on the senses. These images are no more than traces in the brain made by the animal spirits, and as a more forceful current cuts a more defined river into a landscape, in the same fashion, 'we imagine things more strongly in proportion as these traces are deeper and better engraved' by the force and repetition of animal spirits passing through them.[50] If an object of the same type has been perceived previously through the sensory organs, the animal spirits will retrace those traces. The flow of animal spirits over these traces is essentially what triggers the mind's perception of an idea (of apples in general), complemented by sensations of redness, waxiness, shininess, and the other visual attributes, as determined by a natural judgement such as perspective and relation of size to the body.[51] Furthermore, whenever brain traces are involved, sensations must also occur. This combination of pure perception of an abstract idea with particular sensations gives us the means to perceive the apple both as a member of a class of objects (*an* apple) and as a unique world object (*this* apple).

The involvement of the body in perceiving ideas representative of objects in the world necessitates the experience of perceptual sensations alongside pure perceptions. It is due to the internal movements of the body, then, that we experience perceptual sensations at all: this is what Malebranche means when he says that 'the union of soul and body [...] consists primarily of a mutual relation between sensations and motion in the organs'.[52] But what about the perceptual sensations we experience during episodes of imagining? Are they different to those of sensory

---

[48] See Malebranche 1678/1997, 98; Jolley 1995.

[49] Malebranche 1678/1997, 723–724.

[50] Malebranche 1678/1997, 134. Here we can see an example of similarity between Malebranche and Hume: 'A greater force and vivacity in the impression naturally conveys a greater to the related idea; and 'tis on the degrees of force and vivacity, that the belief depends'. (Hume 1739–1740/2000, Vol. 1, 98.) Cf. Hume 1739–1740/2000, Vol. 1, 67.

[51] See Malebranche 1678/1997, 33–36, 43–44.

[52] Malebranche 1678/1997, 20.

perception? How do they rely on the movements of the body if there is nothing impacting upon the sensory organs to trigger the movements of the animal spirits? Malebranche's answers to these questions, detailed in the next section, involve situating the imagination itself within the body.

4.4 The Psycho-Physiology of Imagination

For Malebranche, one of the helpful tools provided by the body is the imagination. While it is often referred to as a faculty of the mind, Malebranche warns us that such talk of faculties is purely a *façon de parler*: in the opening chapter of the *Search* he refers interchangeably to the 'faculties' and 'properties' of the substances. And in the second *Elucidation* he explains that to talk of the mind's faculties is to talk of the functional states of the mind:

> It should not be imagined that the soul's different faculties [...] are entities different from the soul itself. [...] It is really the soul, then, that perceives, and not the understanding conceived as something different from the soul. The same is true of the will; this faculty is but the soul itself insofar as it loves its perfection and happiness, insofar as it wills to be happy, or insofar as [...] it is made capable of loving everything that appears to it to be good.[53]

Likewise with imagination: it is a 'faculty' of the mind insofar as it denotes an epistemological function or process. Its role is to make present to the mind material beings 'when in fact they are absent', which it does 'by forming images of them, as it were, in the brain'[54]—that is, by directing the animal spirits to previously etched brain traces.

One can immediately conceive of situations in which the imagination could be helpful: the Descartes-inspired Malebranche posits geometry as the obvious example to demonstrate this fact. He explains that

> those who begin the study of geometry conceive very quickly the little demonstrations one explains to them [...] because the ideas of square, circle, and so forth, are tied naturally to the traces of the figures they see before their eyes.[55]

In other words, the imagination works much like a sketchpad allowing the mind to dress up an algebraic equation in sensible qualities. Here Malebranche echoes Descartes's account of the imagination in his *Sixth Meditation* (1641):

> When I imagine a triangle, for example, I do not merely understand that it is a figure bounded by three lines but, at the same time, I also see the three lines with my mind's eye as if they were present; and this is what I call imagining.[56]

---

[53] Malebranche 1678/1997, 560. Malebranche again clarifies his position on faculties in his reply to the *First Objection* in the tenth *Elucidation*, 622–624.
[54] Malebranche 1678/1997, 17.
[55] Malebranche 1678/1997, 104.
[56] Descartes 1984–1991, Vol. 2, 50.

And this position is justified by Descartes in his earlier, unfinished methodological text, *Rules for the Direction of the Mind* (1701). Talking of solving geometrical equations, he states:

> The problem should be re-expressed in terms of the real extension of bodies and should be pictured in our imagination entirely by means of bare figures. Thus it will be perceived much more distinctly by our intellect.[57]

Malebranche restates this rule in terms of focus of attention, claiming that the imagination acts as a powerful influence over the animal spirits (when it is correctly controlled), such that

> the mind is made more attentive without a wasteful division of its capacity and is thus remarkably aided in clearly and distinctly perceiving objects, with the result that it is almost always to our advantage to avail ourselves of its help.[58]

So for the purposes of practising geometry, the imagination helps us to understand problems in familiar (though non-truth-providing) forms. An appropriate analogy would be to consider the sensible images of the imagination in crude cartographical terms: an image may be quite helpful in drawing attention to particular aspects of the idea, but only when considered, so to speak, not to scale.[59]

One point should be carefully noted: for Malebranche (for Descartes too, in fact) the imagination is typically restricted to the recombination of previously perceived sensible qualities (and their corresponding ideas). This is a limit of the body: the animal spirits find difficulty in etching new traces in the brain unless they are forced to do so by the violent effects of sensory impressions.[60] In the absence of such forceful impressions, they retrace old brain traces, 'because the animal spirits [find] some resistance in the parts of the brain whence they should pass, and being easily detoured crowd into the deep traces of the ideas that are more familiar to us'.[61] The phenomenological result is that our imaginary perceptions seldom appear as vivid or familiar as our sensory perceptions: an imagined foghorn will not be louder than the song of a bird sitting outside one's window. But a distinction

---

[57] Descartes 1984–1991, Vol. 1, 56; emphasis removed.

[58] Malebranche 1678/1997, 419; see also Malebranche 1678/1997, 132; Sutton 1998a, 115–146.

[59] Of course the imagination can only aid the geometrically inclined mind in relatively simple procedures. Imagining shapes is not a *better* means of practising mathematics than is algebraic geometry: 'With the mind neither hampered nor occupied with having to represent a great many figures and an infinite number of lines, it can thus perceive at a single glance what it could not otherwise see, because the mind can penetrate further and embrace more things when its capacity is used economically' (Malebranche 1678/1997, 209). Malebranche also mentions that it is by way of the pure understanding that we can accurately perceive a figure of a thousand sides (Malebranche 1678/1997, 16), hinting at Descartes' own distinction between the intellect and the imagination (see Descartes 1984–1991, Vol. 2, 50–51). Hume adopts a very similar standpoint towards geometry through sensory perception to the one we find in the two rationalists, though of course he discards the notion of necessary truths perceived in rationalistic ideas; see Hume 1739–1740/2000, Vol. 1, 50–52.

[60] Malebranche 1678/1997, 88.

[61] Malebranche 1678/1997, 135.

through vivacity is far from exact, and at base sensory and imaginary perceptions are of the same epistemological nature. Hence Malebranche's claim that the difference between proper sensory perceptions and our somewhat dimmer imaginary perceptions is one of degree.[62]

It is therefore by way of our minds' embodiment that we are able to perceive worldly objects, either through their impact on the sensory organs resulting in the etching of new brain traces by the animal spirits, or through recollection of their sensory attributes by the retracing of prior etchings. In either case, the movements of the animal spirits in the brain dictate what kinds of perceptual sensations will appear to the mind. But the animal spirits do not exist merely for the mind's perceptions through its body; indeed, they are the body's means of protecting and guiding itself without the authority of the willing mind. The spirits flow throughout the entirety of the body, causing physiological changes which themselves trigger more sensations in the mind. Following Descartes, in particular his *Passions of the Soul* of 1649, Malebranche talks at length of these processes of the passions in Book V of the *Search*. Complementing these passions are the emotional sensations of the mind whose jobs differ from those of the perceptual sensations. The passions and their emotional sensations are explored in the following section.

4.5 Passions and Emotional Sensations

Malebranche oscillates between two uses of the term *passion*: at times, he talks broadly of passions as the movements of the animal spirits within the body alongside the sensations they trigger, while at others, he talks more narrowly of *les passions de l'âme* merely as the sensations or impressions that incline the mind towards loving its body and anything useful to its preservation, with the former use of the term standing as the 'natural or occasional cause of these impressions'.[63] For the sake of clarity, I will maintain a terminological distinction between passions as seven-part psycho-physiological processes,[64] and the sensations that contribute to those processes, referring to each as passions and emotional sensations respectively.

Fully fledged passions are sequential and occurrent: they involve seven different elements, each one leading to the next. The first step is the mind's perception of an object and its relation to us *qua* body-and-mind composites. This causes in the will, secondly, an impulse towards the object if it appears good or an aversion if it appears evil. The third element found in episodes of passions is an accompanying sensation of, say, love, aversion, desire, joy, or sadness. These affective sensations correspond to the fourth element, a redirection of the blood and animal spirits to the 'external'

---

[62] Malebranche 1678/1997, 87. This view echoes through Hume's comparison of impressions and ideas; see Hume 1739–1740/2000, Vol. 1, 7–10; Hume 1748/2007, 15.
[63] Malebranche 1678/1997, 338.
[64] Malebranche 1678/1997, 347–352.

parts of the body such as the facial muscles and limbs. That is, the physiological changes which characterise passions such as anger or joy (frowns or smiles) are the direct causes of a violent rerouting of agitated animal spirits to specific parts of the body correlated with the experience of particular emotional sensations. The fifth element is a feedback sensation from the body to the mind whereby it experiences the flow of agitated animal spirits throughout the body, since the motions of the body and mind are reciprocal. The sixth element comes in the form of another sensation of love, aversion, joy, desire, or sadness caused by disturbances in the brain due to the highly agitated animal spirits.

We should pause here to consider this sixth element. Malebranche's distinction between this particular emotional sensation and that which appears as the third element is important: the sixth element is caused by the animal spirits in the brain rather than by an impulse of the will. It is also likely to be much livelier than an emotional sensation caused by a judgement. As Susan James explains, the 'workings of our "machine", as Malebranche calls it, strengthen our passions, and in doing so heighten our consciousness of harmful or pleasurable states of affairs'.[65] It seems that Malebranche is offering an explanation of our tendencies to overreact when we find ourselves in passionate states: indeed, we certainly cannot claim that men are free from the dominance of the passions.[66]

The seventh and final element of any passion is 'a certain sensation of joy, or rather of inner delight, that fixes the soul in its passion and assures it that it is in the proper state with regard to the object it is considering'.[67] That is, every passion—whether it is one of anger, joy, sadness, or love—will produce a positive emotional sensation because of the fact that it demonstrates the harmony between mind and body. Malebranche cites the pleasure that accompanies sadness evoked by theatrical performances as evidence of this last element: 'this pleasure increases with the sadness, whereas pleasure never increases with pain'.[68] Furthermore, this pleasure will occur even in those cases in which the object of a passion appears to be missing (we might call these 'moods').

The emotional sensations found in episodic passions are similar to the affective perceptual sensations of pleasure and pain, and they should be classed as sensations proper because, like colours and odours, they are ways of perceiving objects (via ideas). Yet they are distinguishable from other sensations in that they go beyond the concerns of the body alone: they also point to the very important relation between body and mind. They are likewise distinguishable as they are always preceded by some judgement on the part of the mind.[69] We can say then that passions hold dual intentionality. On the one hand, their elements typically point out various relations between the object considered and the perceiving agent; on the other, thanks to the

---

[65] James 1997, 113.
[66] Malebranche 1678/1997, 346.
[67] Malebranche 1678/1997, 349.
[68] Malebranche 1678/1997, 201.
[69] Malebranche 1678/1997, 201.

final 'inner delight' felt in each of the passions, they draw our attention to the fact that our minds are very much embodied.

In the next section we will explore the interaction between the imagination and the passions. We will see that, for Malebranche, the passions not only play a significant role in the preservation of the body, but that they are also crucial aspects of social interaction.

4.6 Imagination and the Passions

Passions share many similarities with sensory perception and imagination: all three involve activity of animal spirits and brain traces, and all three correspond with types of perceptions which can be called sensations. Malebranche explains that on the occasions when our animal spirits are unusually active, the two types of perceptions (sensory perception and imagination) can come much closer to one another:

> However, it sometimes happens that persons whose animal spirits are highly agitated by fasting, vigils, a high fever, or some violent passion have the internal fibers of their brains set in motion as forcefully as by external objects. Because of this such people *sense* what they should only *imagine*, and they think they see objects before their eyes, which are only in their imaginations.[70]

Considering that passions are episodes of 'extraordinary motion in the animal spirits',[71] one important implication concerns the influence of the passions and the imagination on one another: if an imagining is adequately vivid, and if its object is something that warrants a response by way of a passion, then we should experience proportionately vivid emotional sensations. This is especially true when imagining objects we conceive of as possessing bodily good, 'for the imagination always increases the ideas of things that we love and that are related to the body'.[72] Equally, if we are suffering a passion and our animal spirits are highly aroused, we will imagine things with much greater force than we would were we in a calmer state. And with respect to the cause of a passion, there is no clear distinction—or reason to distinguish—between a sensed object and an imagined object. In Malebranche's example, a man can experience the same sort of passion, with the same intensity, whether he is insulted by someone or merely imagines being insulted.[73] Malebranche describes in the following story the reaction of the man who has potentially been insulted by another:

> But nature has provided well for him [the victim of insult], for at the prospect of losing a great good, the face naturally takes on aspects of rage and despair so lively and unexpected that they disarm the most impassioned men and, as it were, immobilize them. This terrible and unexpected view of death's trappings painted by the hand of nature on the face of an

---

[70] Malebranche 1678/1997, 88.
[71] Malebranche 1678/1997, 337.
[72] Malebranche 1678/1997, 262.
[73] Malebranche 1678/1997, 349.

unhappy man arrests the vengeance-provoking motion of his enemy's spirits and blood [...]
As a result of this, he [the alleged offender] is mechanically taken by impulses of
compassion, which naturally incline his soul to accede to motives of charity and mercy.[74]

Here two passions are at play, both of which occur for the protection of their bearers' bodies. However, they both also perform strong social roles: the first passion of rage includes elements which communicate something of the pain of the insulted man to the one who potentially insulted him; the second passion of charity and mercy reflects something of the suffering of the insulted man. The result of this clashing of passions is a sort of nullification of each passion. Importantly, all this communication of passions occurs mechanically, likely by a kind of natural judgement; there is no need to explain the episode in terms of deliberation or activity of the will.[75] This point is very important for Malebranche's account of the passions as it bestows upon them the status of natural peacekeepers in cases of social interaction such as this.

This tendency towards sympathy provides the bedrock for something like a natural ethics in a Malebranchean world. In the next section I will explore this notion in more detail. In doing so, I hope to draw out a common thread between Malebranche and Hume: the latter, I will argue, very likely read Malebranche in the sort of way that I have outlined here.

4.7 Natural Ethics and Sympathetic Impulses: Malebranche's Influence on Hume

Malebranche sees the coupling of the imagination and passions as dangerous for the mind: it draws our attention away from the discovery of eternal truths (and the 'true good' of God) and leads us to spend too much time worrying about the body. However, there is reason to believe this is not the whole story: lurking behind the dim warnings of the body's influence over the mind is a more positive account of the operations (and co-operations) of the imagination and the passions. Malebranche has available an alternative to his theocentric moral theory offered in the *Treatise on Ethics* of 1684, though his commitment to theodicy sees him underplay this theme and instead place the onus of moral decision-making solely on the rational mind. Yet Malebranche could perhaps have put this view forward as a sort of consolation, a natural ethics, had he anticipated (and conceded) Hume's treatment of occasionalism as a 'superfluous' account of causation.[76] The consolatory view rests on three

---

[74] Malebranche 1678/1997, 351.

[75] Though Malebranche mentions that it is the will which judges that the utterances perceived by the soon-to-be attacker are insulting, he need not claim that it is the will which triggers the passion itself. Indeed he maintains that it is a judgement *qua* perception that triggers our passions, and perceptions are matters for the understanding, not the will *per se*. See Malebranche 1678/1997, 351.

[76] Hume 1739–1740/2000, Vol. 1, 107–108; see also Hume 1748/2007, 67; Kail 2008b, 55–80.

elements of Malebranche's philosophy. First, the influences of the imagination and the passions on each other largely determine the behaviour of the impassioned, as we saw above. The second element is implied in Malebranche's description of the differences and similarities between particular sensations across different minds. Malebranche tells us that all men have the same nature and share the same perceptions of ideas, thus all persons share a common inclination towards happiness and the avoidance of pain and evil (though we cannot be sure that we all perceive the same sensations in the same way).[77] Finally, Malebranche emphasises that all men experience a natural inclination of friendship or compassion felt towards other men. This inclination of sympathy is always joined to the passions and is the strongest natural union found between God's works.[78] It is a concern of the embodied mind rather than the pure intellect:

> This compassion in bodies produces a compassion in the spirits. It excites us to help others because in so doing we help ourselves. Finally, it checks our malice and cruelty. For the horror of blood, the fear of death—in a word, the sensible impression of compassion—often prevents the massacre of animals, even by those most convinced that they are merely machines, because most men are unable to kill them without themselves being wounded by the counterblow of compassion.[79]

That we share common passions and sympathise with other creatures (human or animal) is evidenced in the various perceivable modifications of the bodies (particularly the faces) of those with whom we interact, as we saw in the previous section.[80] Malebranche has no reason to claim that this sort of sympathetic reaction is restricted to real-life situations; comparable imagined situations could just as easily conjure passions in the imaginer. Thus we can feasibly utilise the imagination *in order* to stir up passions of joy or sadness so that they can operate as immediate feedback systems with respect to the scenarios we imagine. If we are faced with a situation in which our actions will affect other persons, we can use the affective feedback afforded by the forward-thinking imagination in order to assess the best course of action by first imagining the situation from our own point of view (to assess the impact of our affect) and then from the other participants' points of view (to assess the impact of their affect). This latter imagining will provide us with passions corresponding to the imagined actions that we can appropriately label empathetic or sympathetic. We can thus judge our actions based on whether or not we are causing other persons bodily pleasures or pains: since negative passions are to be avoided, we recognise that we should not cause them in others. (The 'should'

---

[77] Malebranche 1678/1997, 238–239.

[78] Malebranche 1678/1997, 330–331. See also James 2005 for an account of Hume's appropriation of Malebranche.

[79] Malebranche 1678/1997, 114.

[80] At least, we are all disposed to respond to the effects of a particular passion in the same manner. The *actual* phenomenological quality of sadness, say, may differ between minds; this is a question we could never resolve given that we do not have access to each other's phenomenological experiences. While we cannot be sure that sensations between minds are phenomenologically equivalent, we can at best be confident that they are *functionally* equivalent. See Malebranche 1678/1997, 63–66, 238–239.

here is naturalistic, just as a dryness of the throat informs us that we should drink.) Through a Malebranchean account of passions and imagination (and by focusing on the contingent 'truths' of Malebranchean science rather than his metaphysics), we can lay the groundwork for a world- and especially body-centric ethical system which relies not on moral duty through knowledge of God, but on the complex relations and interactions between minds and bodies. Such a theory would rely on the will only insofar as it seeks pleasure and avoids pain: any judgements made would be guided by one's natural inclination towards compassion, rather than by reflection on the truths afforded by pure perceptions of eternal ideas. In short, exercising the moral imagination would mean acting in a morally sound way without having to reflect on one's duty to God.

Readers familiar with Hume's works will likely notice strong similarities between the above account of what I have been calling a natural ethics through Malebranchean passions and the foundations of the ethics of *le bon David*. In Book 3 of his first major work, *A Treatise of Human Nature* (1739–1740), Hume explains that 'vice and virtue are not discoverable merely by reason'; rather they are differentiated according to some 'sentiment they occasion'.[81] Morality is subsequently 'more properly felt than judg'd of'.[82] It is precisely the sentiment triggered by the perception of an event that sees us determine its moral valence: 'An action, or sentiment, or character is virtuous or vicious; why? because its view causes a pleasure or uneasiness of a particular kind'.[83] Thus, we deem something to be virtuous or vicious by appealing to the particular sentiment we experience in perceiving that something. What I hope to have demonstrated in this paper is that the Malebranchean embodied mind, too, has the capacity to, and often does, experience the 'particular kind' of sentiment that Hume employs as the bedrock of his moral theory.[84] Granted, Malebranche does not see much of moral value in such sentiments, given that his theologically informed moral theory relies on eternal truths. But Hume, in denying that the mind has access to some such intellectual realm, instead relies upon the scientific 'truths' that come from examining these sentiments. Hume, then, adopts Malebranchean science, in part at least, to replace Malebranchean metaphysics as the means of grounding a moral theory.

We can push the connection further by noting that Malebranche's theory of embodied passions is highly compatible with Hume's sceptical materialism,[85] and that Hume saw in Malebranche many of the metaphysical resources from which to build his own theory. Hume was certainly no stranger to Malebranche's philosophy. Complementing the facts that numerous sections of the *Treatise* contain near

---

[81] Hume 1739–1740/2000, Vol. 1, 302.

[82] Hume 1739–1740/2000, Vol. 1, 302.

[83] Hume 1739–1740/2000, Vol. 1, 303.

[84] For more detailed accounts of Hume's moral philosophy see Book 3 of Hume 1739–1740/2000, Vol. 1. Many ideas therein have been reworked and published in Hume 1751/1998. Important discussions include Capaldi 1989; Cohon 2008; Loeb 1977; Mackie 1980.

[85] Stephen Buckle argues that Hume's account of the passions is 'implicitly materialist'. See Buckle 2012, 204.

word-for-word appropriations of the *Search*,[86] and that Hume places that work at the head of a list of philosophical texts recommended to his friend Michael Ramsay,[87] several recent scholars have exposed common themes between the two philosophers' works.[88] While much attention has been paid to Hume's adaptation of Malebranche's criticisms of orthodox Cartesian claims that the mind is better known than the body, or to the shift from Malebranche's occasionalism to Hume's sceptical views on causation, few have attempted to expose any great debt Hume owes to Malebranche's metaphysically informed moral theory. And this is not without reason: Hume's project is largely anti-metaphysical. It is quite likely, however, that Hume noticed in Malebranche something akin to the scientifically guided natural ethics described above; indeed, as Charles McCracken points out, 'Hume had a gift for seeing in the ideas of others possibilities that were not always apparent to their originators'.[89] In short, the foundations of Hume's moral theory are readily available within Malebranche's psycho-physiological system.

McCracken notes the significance of the Malebranchean passions and their emotional sensations for acting in a morally sound way: it was our pre-lapsarian default, as it were. He also explains that this theory of moral action through the passions is very similar to the one arrived at by Hume: 'Before their fall, it seems, our first parents were good Humeans, distinguishing right from wrong by immediate *sentiment*; only in the day in which something went awry in the Garden did they have to begin to *reason* about morality'.[90] Similarly, Peter Kail observes that 'the Christian Platonist, as McCracken termed Malebranche, collapses into an empiricist naturalist when the intellect is decapitated'.[91] (Again, this is also true of Malebranche's science.) While it is certainly the case that Hume takes aim at Malebranche not just in the *Treatise*, but also in the first *Enquiry*, most of the Scot's criticisms of the Frenchman specifically target problems arising due to the latter's theological commitments. By utilising numerous Malebranchean elements to his own ends, Hume is able to bring the battle to Malebranche's field. Take God out of Malebranche's philosophical system and it will certainly start to fall apart: since man is not a light unto himself, as Malebranche repeatedly preaches throughout his works, his search after truth will fail miserably if he attempts it without the illumination of eternal ideas through his union with his Author. But what remains is not reduced to mere rigid mechanism. The Malebranchean embodied mind maintains full access to its sensations—both perceptual and emotional—and can enjoy social interactions (though arguably in a much less sophisticated sense than previously) with its peers.

Voltaire, then, was too harsh in his dismissive treatment of Malebranche. It is simply not the case that we see all things in God: the Malebranchean mind has

---

[86] See McCracken 1983, 257–261; Kail 2008b.
[87] Hume's letter to Ramsay can be found in Hume 1748/2007, 203–204.
[88] More prominent studies on the various Malebranche-Hume connections include McCracken 1983; James 2005; Kail 2008a, b; Buckle 2001; Gaukroger 2010, 439–440.
[89] McCracken 1983, 255. Cf. Buckle 2001, 192.
[90] McCracken 1983, 286.
[91] Kail 2008b, 76.

available a rich repertoire of passions and affective sensations by virtue not of its union with its Author, but of its union with its body. Far from a perfect philosophical system, Malebranche's theory is strong enough to have merited careful consideration from one of the eighteenth century's most important thinkers. That alone is enough to justify Malebranche's growing reputation as a highly influential figure in the discourse of sensibility in the Enlightenment era.

Acknowledgments Versions of this paper were presented at the *Sensibilité* conference in Brisbane, December 2010, and at the inaugural conference of the Centre for the History of Philosophy in York, UK, May 2011. I wish to thank all audience members who provided feedback. Thanks also to Stephen Buckle, Devin Curry, Karen Detlefsen, Daniel Garber, Stephen Gaukroger, Nabeel Hamid, Martyn Lloyd, Verónica Muriel, and Alison Simmons for helpful discussions, and most importantly to John Sutton, whose prompts and comments have proved indispensable.

References

Buckle, Stephen. 2001. *Hume's enlightenment tract: The unity and purpose of an enquiry concerning human understanding*. Oxford/New York: Oxford University Press.
Buckle, Stephen. 2012. Hume on the passions. *Philosophy* 87: 189–213.
Capaldi, Nicholas. 1989. *Hume's place in moral philosophy*. New York: Peter Lang.
Cohon, Rachel. 2008. *Hume's morality: Feeling and fabrication*. Oxford/New York: Oxford University Press.
Descartes, René. 1984–1991. *The Philosophical Writings of Descartes*, 3 vols. Trans. and eds. J. Cottingham, R. Stoothoff, D. Murdoch, and A. Kenny. Cambridge: Cambridge University Press.
Gaukroger, Stephen. 2010. *The collapse of mechanism and the rise of sensibility: Science and the shaping of modernity, 1680–1760*. Oxford: Oxford University Press.
Hume, David. 1739–1740/2000. *A treatise of human nature*, 2 vols, ed. David F. Norton and Mary J. Norton. Oxford: Oxford University Press.
Hume, David. 1748/2007. *An enquiry concerning human understanding and other writings*, ed. S. Buckle. Cambridge: Cambridge University Press.
Hume, David. 1751/1998. *An enquiry concerning the principles of morals*, ed. Tom L. Beauchamp. Oxford: Oxford University Press.
James, Susan. 1997. *Passion and action: The emotions in seventeenth-century philosophy*. Oxford/New York: Oxford University Press.
James, Susan. 2005. Sympathy and comparison: Two principles of human nature. In *Impressions of Hume*, ed. Marina Frasca-Spada and P.J.E. Kail, 107–124. Oxford: Oxford University Press.
Jolley, Nicholas. 1988. Leibniz and Malebranche on innate ideas. *The Philosophical Review* 97: 71–91.
Jolley, Nicholas. 1994. Intellect and illumination in Malebranche. *Journal of the History of Philosophy* 32: 209–224.
Jolley, Nicholas. 1995. Sensation, intentionality, and animal consciousness: Malebranche's theory of the mind. *Ratio* 2: 128–142.
Kail, Peter J.E. 2008a. Hume, Malebranche, and 'Rationalism'. *Philosophy* 83: 311–332.
Kail, Peter J.E. 2008b. On Hume's appropriation of Malebranche: Causation and self. *European Journal of Philosophy* 16: 55–80.
Loeb, Louis. 1977. Hume's moral sentiments and the structure of the treatise. *Journal of the History of Philosophy* 15: 395–403.
Mackie, J.L. 1980. *Hume's moral theory*. London/New York: Routledge.
Malebranche, Nicolas. 1678/1997. *The Search after Truth*. Trans. and ed. Thomas M. Lennon and Paul J. Olscamp. Cambridge: Cambridge University Press.

McCracken, Charles. 1983. *Malebranche and British philosophy*. Oxford: Oxford University Press.
Nadler, Steven. 1992. *Malebranche and ideas*. New York: Oxford University Press.
Nadler, Steven. 2011. *Occasionalism: Causation among the Cartesians*. Oxford/New York: Oxford University Press.
Schmaltz, Tad M. 1996. *Malebranche's theory of the soul: A Cartesian interpretation*. New York/Oxford: Oxford University Press.
Simmons, Alison. 2009. Sensations in a Malebranchean mind. In *Topics in early modern philosophy of mind*, ed. J. Miller, 105–129. Dordrecht: Springer.
Sutton, John. 1998a. Controlling the passions: Passion, memory, and the moral physiology of self in seventeenth-century neurophilosophy. In *The soft underbelly of reason*, ed. Stephen Gaukroger, 115–146. London: Routledge.
Sutton, John. 1998b. *Philosophy and memory traces: Descartes to connectionism*. Cambridge: Cambridge University Press.
Voltaire. [François Marie Arouet]. 1733/2003. *Philosophical Letters: Letters Concerning the English Nation*. Trans. and ed. Ernest Dilworth. Mineola: Dover.

Chapter 5
Feeling Better: Moral Sense and Sensibility in Enlightenment Thought

Alexander Cook

Abstract For much of the eighteenth century in Europe, the concept of 'sensibility' formed a bridge between nature and culture, and between the body and the moral world. In doing so, it provided a site for the emergence of a range of projects in what this chapter labels affective pedagogy—techniques for training the unruly passions and for nurturing social sentiments as an inoculation against them. Though such pedagogy could take various forms, ranging from sentimental novels to dietetic regimes, it was characterised above all by a common attempt to help humanity, quite literally, to *feel* better. By exploring the social and philosophical dynamics that promoted the turn to affective pedagogy, and the anxieties that shaped its manifestations in Britain and France, this chapter illuminates how eighteenth-century understandings of the character and history of emotions were bound up with central moral and political questions of the era.

The eighteenth century in Western Europe has long been characterised as an 'Age of Reason'. Yet despite, or in part because of, that period's preoccupation with reason, it was a time when many people developed an urgent and earnest interest in the limits of human rationality. As such, it was an era that witnessed a striking and in many ways novel concern with the world of emotion. From the rise of the 'sentimental novel', typified by Richardson and Rousseau, across the realms of aesthetics and moral philosophy, to a burgeoning medical literature on the physiology of feeling, this period saw many attempts to provide an anatomy of human sentiment that would explain its origins, its character, and its role in human life. The consequence was a

A. Cook (✉)
School of History and Research School of Social Sciences,
Australian National University, 0200 Canberra, ACT, Australia
e-mail: Alexander.Cook@anu.edu.au

series of passionate debates about the nature and implications of 'sensibility' that crossed discursive genres and national boundaries, tapping into a range of hopes and fears about contemporary European civilisation.

The 'cult of sensibility', as it has often been called, was a diverse phenomenon. Predictably, it has been characterised in various ways. For some scholars, it is associated with a kind of revolt against the narrow rationalism of mainstream enlightenment thought—a proto-romantic ethos linked to an elevation of the intuitive and the natural against a world of calculation, artifice, and social conformity.[1] Others have seen it as a product of emergent commercial society—a bourgeois philosophy articulating a value system designed to challenge or supplant the codes of honour and pedigree associated with a declining feudal order, a vehicle for facilitating urban sociability, the civilising process and even modern patterns of consumption.[2] The culture of sensibility has been linked both to materialist attempts to by-pass the soul and to the neo-pietist religious longings of the late-eighteenth and early nineteenth centuries.[3] Many have highlighted the role of sensibility in period gender politics, its common association with women, and its consequent function in attempts both to validate women's contribution to society and to circumscribe the forms of that contribution.[4]

The diversity of this scholarship can be intimidating but it should not surprise us. We should expect many histories of a concept which the persecuted French materialist philosopher Claude-Adrien Helvétius sought to make the basis of a moral science and which the conservative British Evangelical Hannah More declared to be 'virtue's precious seed'.[5] Sensibility in the eighteenth century was one of those protean terms whose intellectual prestige and fundamental ambiguities invited myriad forms of appropriation.

This chapter is an attempt to shed some light on both that prestige and those ambiguities as they were manifested within the realm of political and moral philosophy. It does not pretend to offer anything like a full explanation for the rise of the culture of sensibility in eighteenth-century Europe, or for the various factors that shaped its local forms. Instead, it seeks, more modestly, to examine some of the dynamics which drove the growing preoccupation with sensibility amongst philosophers and to show how those dynamics can help to explain the central, but contested, status of the concept—both within philosophical discourse and, to some extent, within period social life. The chapter focuses, in particular, on the way in which over the course of the eighteenth century, the concept of sensibility formed a bridge between nature and culture and between the body and the moral world. In doing so, it provided a site for the emergence of a range of projects in what I have called affective pedagogy—techniques for training the unruly passions, for

---

[1] Mornet 1929/1969, 209–229.

[2] Langford 1989, 461–66; Campbell 1983, 279–296.

[3] On the materialist associations of discourse about sensibility in France, in particular, see Vila 1998. On links between sensibility, Methodism, and new forms of pietism in the Anglophone world, see Barker-Benfield 1992, 266–279.

[4] For Britain, see Barker-Benfield 1992. For France, see Steinbrügge 1995.

[5] Helvétius 1770/1989, Vol. 2, 171–195; More 1782/1785, 282.

nurturing social sentiments as an inoculation against other kinds. Though it could take various forms, this affective pedagogy was characterised above all by the attempt to help humanity, quite literally, to feel better. The story told in this chapter is centred on Britain and France. These were not the only countries where issues of sensibility became entangled in moral philosophy.[6] But it is in these countries that debate around these issues was most fully developed and that the contested politics of sensibility seems to have had the most direct impact upon society as a whole.

In its simplest form, the concept of 'sensibility' in the eighteenth century denoted the capacity of a physical organism to register impressions from the external world via the medium of the senses. The body registered the elemental qualities of objects such as hot, cold, rough, smooth, as well as corporeal responses to those objects, such as pleasure or pain, and transferred them to the brain (or the *commune sensorium*, as the relevant part was sometimes called) via the medium of the nerves. There they were combined, processed, and reflected upon to produce our ideas about the world. Early eighteenth-century understandings of this process in both Britain and France were strongly influenced by the work of the English philosopher John Locke, whose theorisation of the fundamental role of the senses in providing human beings with their knowledge and their basic inclinations was crucial to eighteenth-century debates on everything from epistemology to psychology.[7]

This relatively simple model of sensibility continued to have meaning throughout the eighteenth century. Johnson's *Dictionary of the English Language* (1755) defined sensibility as pertaining to 'quickness of sensation' or 'quickness of perception'.[8] Sheridan's *New and Complete Dictionary* (1797) defined it in the same way.[9] Over the course of the century, however, the physiological understanding of sensibility became more complex as a range of anatomical theorists and medical practitioners attempted to model the process in more detailed ways. The work of Albrecht Von Haller was important in popularising the concept of fibres to link the nerve endings to the *commune sensorium*.[10] The influential Montpellier School of medicine in France would play a crucial role in developing period understandings of the physiology of sensibility, while linking it to a broader vitalist cosmology.[11] At the same time, sensibility began to acquire a range of more extensive semantic connotations. For many, the term came to refer not just to elementary physiological responses to the material world, but to an array of relatively complex sentiments relating to that world. In particular, the concept was increasingly linked with notions of aesthetic taste and with emotional responsiveness to the plight of

---

[6] On early German debate about the role of sentiment in moral theory in the early eighteenth century, and its links to both pietism and German 'rationalism', see Norton 1995, Chap. 2.

[7] Barker-Benfield 1992, xvii.

[8] Johnson 1755/1768.

[9] Sheridan 1797.

[10] Haller 1756–1760.

[11] One of its major figures, Théophile de Bordeu, introduced a theory of glandular sensibility that had considerable period prestige. Bordeu served as the *médecin-philosophe* in Denis Diderot's *Rêve de d'Alembert* (composed 1769). See Bordeu 1751; Haigh 1976. On the wider theorisation of the physiology of sensibility through the middle of the eighteenth century, see Vila 1998, 13–80.

other sentient beings, whether that plight was real or imagined via the medium of artistic representation. Thus, by the second half of the eighteenth century, we find the narrator of Lawrence Sterne's *Sentimental Journey through France and Italy* (1768) suggesting that it is entirely owing to 'sensibility' that 'I feel some generous joys and generous cares beyond myself'.[12] In the French *Encyclopédie méthodique* (1789), we find the statement that 'you cannot have either humanity or generosity without sensibility'.[13] Throughout the eighteenth century, the relationship between corporeal sensibility and what we might call imaginative sensibility remained an open question. Many believed there must be some relationship, but where some believed that the two must march together, others suspected they might exist in some kind of tension.

Both conceptions of sensibility played a role in eighteenth-century moral philosophy. To understand how, it is important to have some sense of context. The pre-history of this process extends to a time well before Locke. Indeed, despite Locke's importance to the history of sensibility in epistemology, his moral philosophy impinged relatively little on these debates. The crucial moment seems to have been a point, around the middle of the seventeenth century, when the perceived role of the passions in human life began to be re-cast by a steady stream of philosophers, theologians, and public moralists. Throughout the Middle Ages and the Renaissance, the mainstream of European moral thought worked with models in which the individual was seen as the site of a battle between the spirit and the flesh, or between reason and the appetites, or between virtue and self-interest. Speaking crudely, we can say that it was the goal of moral theory to nurture the former and to suppress the latter term in each pairing. It was the goal of political management to do the same thing or, if it could not manage this, to render the negative impulses socially harmless. The consequence, for political thought, was a preponderance of theories designed either to inculcate virtue in the citizenry by demonising self-interest, or to manage the horde of sinful creatures by the imposition of sovereign law from above.[14]

During the early modern period in western Europe, and particularly during the seventeenth century, this pattern began to change. In politics, in theology, and in moral theory, the status of the body began to rise. Various factors seem to have contributed to this change. The experience of religious war spawned a search for forms of social theory perceived to be independent of confessional theology.[15] The burgeoning status of the natural sciences encouraged the application of naturalistic models to human life. An increasing role for commerce, foreign trade, and the production of luxury goods in the national economies of north-western Europe invited

---

[12] Sterne 1768/2006, 162–163.

[13] Lacratelle 1789, 183. All translations from French texts in this chapter have been made by the author except where otherwise indicated.

[14] See, for example, Nederman 1988; Pocock 1975. More broadly, see Skinner, Quentin 1978, Vol. 2, 148–165.

[15] On the role of the religious wars in fostering this change, see Hirschman 1977, 129–130.

reconsideration of the psychological and social mechanisms which sustained them.[16] The precise causal process is complex and open to dispute. One consequence of these developments, however, was the emergence of a loose tradition of philosophical thought that was anchored in the idea that human beings are driven by their passions rather than by reason. The theory gave rise to forms of thought which suggested that philosophy, morality, and politics should work with, rather than against, the natural inclinations of human beings. It was important, they suggested, to think about the means by which the pursuit of personal inclinations might be made consonant with social order and collective prosperity. In one form, this produced a species of hard-headed political philosophy exemplified by figures such as the much-detested theorist of political order Thomas Hobbes, or the controversial early-eighteenth-century economic thinker Bernard Mandeville, or by someone like Helvétius in France. For thinkers of this kind, the natural aims of human life were conceived as the pursuit of physical pleasure and the avoidance of physical pain. The task of philosophers was to show how individuals could maximise the one and minimise the other in a manner conducive to collective welfare. Theorists in this tradition could have very different conceptions of what this process entailed. For Hobbes, the key goal was to end the war of all against all which existed in the pre-political state, by the establishment of absolute power in the sovereign.[17] For Mandeville, it was to explore the apparent magic by which the selfish pursuit of personal gain seemed to produce the benefits of corporate prosperity and technical advance.[18] For Helvétius, it was to develop a politico-moral science that would teach governments to allocate material rewards to socially desirable behaviour. For all these thinkers, however, the pursuit of corporeal pleasure and the avoidance of corporeal pain remained the primary mechanism of human psychology. At the level of human motivation, their systems were anchored in what Helvétius called 'physical sensibility'.[19]

From the time of Hobbes, however, many were deeply concerned about the implications of a philosophy that rendered self-preservation, or self-interest, the basis of social life and which found the anchor of order to be prudence. Indeed, with the possible exception of the demonised 'atheist' Baruch Spinoza, Hobbes, Mandeville, and Helvétius were probably the most widely attacked philosophers of the period from 1650 to 1790. They were attacked not just by those who defended scriptural conceptions of moral law or conventional distrust of the human body. They were also attacked by many who accepted the basic contention that human beings were moved to action by the passions rather than reason. For critics of this kind, the 'selfish philosophy', as it was pejoratively called, was both inaccurate as a description of human psychology and inappropriate as a basis for moral philosophy. Thus, from the second half of the seventeenth century in England, the so-called 'Cambridge Platonists', a group of Latitudinarian philosopher-theologians such as

---

[16] On these economic processes, and some of their cultural effects, see Brewer and Porter 1993; Berg and Eger 2003.
[17] Hobbes 1651/1985, Chap. 13.
[18] Mandeville 1714/1989.
[19] Helvétius 1770/1989, Vol. 2, 569.

Henry More and Richard Cumberland, argued against Hobbes that human beings had an in-built capacity for sympathy and a natural impulse towards benevolence that existed alongside the urge for self-preservation. This capacity was part of a Providential plan for human community and the key underpinning of both sociability and morality. It was not a matter of rational recognition of moral duty. For More, 'these Natural and Radical Affections' were 'antecedent to all notion and cogitation'. They were 'a sort of confused muttering or whispering [...] of Divine law'.[20] In France at the same time, a similar campaign to defend the passions by insisting on their benevolent and Providential character was launched by the Augustinian monk Nicolas Malebranche, whose thought is explored elsewhere in this volume.

It is important to recognise that the seventeenth century did not invent the philosophical theory that human morality was grounded in the passions rather than reason—or that sympathy played a special role in it. The theory had classical origins of which contemporaries were well aware. A key participant in this debate, Adam Smith, would later attribute the origins of what he called the 'benevolent system' to the eclectic philosophers of Rome after the Age of Augustus.[21] It was in the wake of Hobbes, however, and in the wider context of a perceived breakdown of the traditional sources of moral authority, that the challenge of finding a workable natural basis for morality crystallised for the moderns.[22]

The work of thinkers like the Cambridge Platonists has frequently been cited as a key precursor to the rising culture of sensibility in the eighteenth century.[23] It is striking, however, that the word 'sensibility' was infrequently used in these late-seventeenth-century moral debates. One finds it rarely in the work of the Cambridge Platonists or in Malebranche. When it does occur, it carries less baggage than the term would later acquire. It is only over the course of the succeeding century that the language of benevolence and sympathy in moral philosophy began to become more intimately entangled with the evolving physiological and aesthetic language of sensibility. The process by which it did is instructive.

In the early part of the eighteenth century, the most important and influential thinkers to attack the Hobbesian psychological model from a perspective supportive of the passions were those who have subsequently been labelled the school of 'moral sense'. This 'school', which was primarily British in its classical manifestation, although its influence extended widely across the channel, includes among its more eminent representatives, Anthony Ashley Cooper the Earl of Shaftesbury, the Scots-Irish philosopher Francis Hutcheson, and, in more complicated ways, those doyens

---

[20] More, Henry. 1690. *Enchiridion Ethicum: The English Translation of 1690*. Latin edition: 1656. Cited in Fiering 1976, 199.

[21] Smith 1759/2002, 354.

[22] In Smith's eyes, it was the 'odious' doctrine of Hobbes that had really created the need, though he accorded also a special place to Mandeville amongst the great enemies of modern moral thought. Smith 1759/2002, 376, 363–364.

[23] On the links between the culture of sensibility and the thought of the late-seventeenth-century 'Cambridge Platonists', see Crane 1934, 205–230. For a contrary view, seeking the origins of sensibility in British low-church culture, see Greene 1977. For an attempted arbitration, see Bruyn 1981.

of the Scottish Enlightenment David Hume and Adam Smith. The term 'moral sense' is derived from Shaftesbury.[24] Its first systematic explanation comes in the work of Francis Hutcheson, who published a series of important tracts in the 1720s, the most famous of which is an *Inquiry into the Original of our Ideas of Beauty and Virtue* (1725).[25] Its title, which joins questions of aesthetics to questions of morality, gives us an important clue about one of the channels through which, for many in the eighteenth century, questions of 'morality' became linked to questions of 'sensibility' in the extensive imaginative sense. Hutcheson argued that, in addition to Locke's account of the external senses, there were a number of 'internal senses'—inbuilt capacities to respond with emotional approbation or hostility to phenomena such as beauty or virtue, ugliness or vice. These were not, Hutcheson was keen to point out, innate ideas. He accepted, if perhaps under slight duress, Locke's rejection of such phenomena.[26] Yet neither were they products of the Lockean process of reflection on sensory inputs. They were quasi-immediate and quasi-corporeal reactions to the world. For moral purposes, the most important of these was sympathy—an innate capacity to feel the pleasures and pains of others.

There were significant differences among thinkers in this tradition regarding the implications of a passionate account of morality, but for all of them, it was a matter of central importance to establish that the inclination to feel sympathy with others was natural in all human beings.[27] As Hutcheson claimed:

> The AUTHOR of Nature has much better furnish'd us for a virtuous Conduct, than our Moralists seem to imagine, by almost as quick and powerful Instructions, as we have for the preservation of our Bodys. He has made *Virtue a lovely Form*, to excite our pursuit of it; and has given us *strong Affections* to be the Springs of each virtuous Action.[28]

As Hutcheson's language makes clear, this response to virtue was conceived as a quasi-aesthetic response. He later elaborated the point:

> As the author of nature has determin'd us to receive, by our *external senses*, pleasant or disagreeable Ideas of Objects according as they are useful or hurtful to our Bodies; and to receive from *uniform Objects* the Pleasures of *Beauty* and *Harmony*, to excite us to the Pursuit of Knowledge [...] so he has given us a Moral Sense to direct our Actions, and to give us still *nobler Pleasures*; so that while we are only intending the Good of others, we undesignedly promote our own greatest *private Good*.[29]

In structural terms, we can see that this project is very similar to that of those theorists who sought to accommodate moral philosophy to a psychology of self-interest. It was an attempt to link the social and the natural orders. Hutcheson himself recognised this relationship to theories of pleasure-drive psychology:

---

[24] Voitle 1955. On the broader tradition, see Darwall 1995.
[25] Hutcheson 1726.
[26] For Hutcheson's ambivalent views on the Lockean conception of innate ideas, see Carey 1997, 281.
[27] On the differences, see Radcliffe 2004.
[28] Hutcheson 1726, xv.
[29] Hutcheson 1726, 134.

> It may perhaps seem strange, that when in this *Treatise* Virtue is suppos'd *disinterested*; yet so much Pains [sic] is taken […] to prove the *Pleasures* of *Virtue* to be the greatest we are capable of, and that consequently it is our truest *Interest* to be *virtuous*.[30]

The difference between Hutcheson and someone like Helvétius is essentially the former's strong emphasis on 'sympathy' as a biological predisposition necessary to anchor this connection between virtue and pleasure, and a distaste for the language of self-interest in moral philosophy. As befits a moral philosophy linked so closely to aesthetics, the concept of 'taste' would come to play a central role.

For Hutcheson, this 'moral sense' was part of a divinely sanctioned order that operated to ensure the compatibility of personal satisfaction with social harmony. It was driven, in part, by a belief that a benevolent God would not ground public prosperity in private frustration. For others, such as Hume, this theological anchoring seems to have been less important. As he put it in his *Enquiry Concerning the Principles of Morals*:

> It is needless to push our researches so far as to ask, Why we have humanity or a fellow-feeling for others. It is sufficient that this is experienced to be a principle in human nature.[31]

Indeed, Hume was prone, on occasion, to expose the potentially relativistic implications of an account of morality that anchored it in emotional disposition. His claim that 'tis not contrary to reason to prefer the ruin of the whole world to the scratching of my finger' is enough to disturb any reader.[32] But he retained enough confidence in the pervasive benevolence of humanity to believe this situation was likely to be rare. It was not, however, logically impossible. And that was potentially worrying. It meant that moral theory of this kind had a huge investment in an optimistic characterisation of human nature. In his *Theory of Moral Sentiments*, published in 1759, Adam Smith reaffirmed the universal character of benevolence:

> How selfish soever man may be supposed, there are evidently some principles in his nature, which interest him in the fortune of others, and render their happiness necessary to him, though he derives nothing from it except the pleasure of seeing it. This sentiment, like all the other original passions of human nature, is by no means confined to the virtuous and the humane, though they perhaps may feel it with the most exquisite sensibility.[33]

This passage, which opens Smith's key work of moral theory, is clearly designed to affirm the universal character of moral sentiment. But that final clause is important. It gives a hint of the manner in which the concept of sensibility began to function within the philosophy of moral sense. It was tied to a rhetoric of exhortation designed to promote what Hutcheson referred to as 'nobler pleasures'.[34]

Despite this insistence on the universal capacity for sympathetic sensibility amongst the theorists of moral sense, most thinkers in the eighteenth century

---

[30] Hutcheson 1728, vi.

[31] Hume 1739–1740/2006, 223.

[32] Hume 1739–1740/2006, 62.

[33] Smith 1759/2002, 11.

[34] On the social and philosophical politics of defining pleasure in the eighteenth century, see Cook 2009. See also more generally, Kaiser 2010.

accepted that sensibility was distributed through the population in variable quantities. Yet it remained a matter of dispute exactly how that distribution occurred. For some, it was largely a question of natural variation. Alexander Gerard, a professor of moral philosophy in Aberdeen and the author of an influential *Essay on Taste* published in 1759, argued 'sensibility very much depends on the *original* construction of the mind; it being, less than any other of the qualities of good taste, *improveable* by use'.[35] The eminent Scottish medic George Cheyne (1671–1743), doctor to the novelist Samuel Richardson and a key theorist of sensibility in the Anglophone world, claimed in 1733 that 'there are as many and as different Degrees of *Sensibility* or of *Feeling* as there are Degrees of *Intelligence* and Perception in *human* Creatures'.[36] Cheyne claimed, like Gerard, that sensibility was distributed in large part by nature:

> As none have it in their Option to choose for themselves their own particular *Frame* of Mind nor *Constitution* of Body; so none can choose *his* own Degree of *Sensibility*. That is given him by the *Author* of his *Nature*, and is already determined.[37]

Cheyne believed, nonetheless, that variations in sensibility were not randomly distributed and could loosely be attributed to groups. The title of his major work, *The English Malady*, gives some indication of a key aspect of his thoughts on the matter. The English, he felt, were particularly given to a certain physiological derangement of their sensibility related to collective lifestyle. He was not alone in seeking a taxonomy of sensibility according to regional type. In a century much given to speculation on the effects of climate upon human character, it was sometimes seen as a key factor in the distribution of sentiment. Montesquieu, perhaps the century's most influential proponent of a climate-based anthropology, suggested that:

> In cold countries, one will have little sensibility for pleasures; it will be greater in temperate countries; in hot countries it will be extreme. As one distinguishes climates by degrees of latitude, one could also distinguish them, so to speak, by degrees of sensibility.[38]

Montesquieu was not, it should be added, amongst those who sought to make sensibility serve a systematic role in moral philosophy. Indeed, Montesquieu's cartography of hedonistic responsiveness was inversely related to some of the other human capacities that contemporaries attributed to sensibility. For him, 'as you approach the southern regions (*les pays du midi*), you will believe yourself distanced from morality itself' because 'more lively passions will multiply crimes'.[39] Yet for those who did believe there was a link between sensibility and morality, the question of how that link operated within individuals and across populations remained a source of debate throughout the century.

If some linked sensibility to geography or ethnicity, others linked it to gender. Much has been written on the particular association of women with sensibility, particularly in the second half of the eighteenth century. This association can be found

---

[35] Gerard 1759/1764, 100.
[36] Cheyne 1733/1991, 366.
[37] Cheyne 1733/1991, 366.
[38] Montesquieu 1748/1768, Vol. 2, 35–36.
[39] Montesquieu 1748/1768, Vol. 2, 37.

both amongst those who prized the quality and amongst those who did not. For critics of the culture of sensibility, it was a constant leitmotif to associate it with the sighs of women. Its perceived spread across society was seen by the more austere thinkers of the age to signify a rising tide of effeminacy and a possible portent of military destruction. For some, the perceived responsiveness of women to both their physical environment and the distress of others was a sign of women's providential role as the moral guardians of society. For others, it was a sign of weakness and incapacity to develop as fully rational beings.[40] The defenders of sensibility, predictably, sought to break its association with both effeminacy and irrationality. But when Hannah More praised 'the tender moralist of Tweed', the novelist Henry McKenzie, by suggesting that 'your *Man of Feeling* is a man indeed', it showed a defensive rejection of a lingering social suspicion.[41]

For many, the distribution of sensibility was, in part, a question of class. This was true not just in relation to that sensibility that enabled its possessors to respond to the charms of higher culture, but even to forms of moral sentiment. The *Encyclopaedia Britannica* claimed in 1798 that 'in all places, the vulgar have little of the sympathy of polished bosoms'.[42] Philosophers, too, subscribed to this notion. Despite his commitment to universal benevolence, Adam Smith believed 'the amiable virtue of humanity requires surely a sensibility, much beyond what is possessed by the rude vulgar of mankind'.[43]

These widespread acknowledgements, or assertions, that sensibility was distributed unevenly highlight one of the paradoxes of the concept's history in our period. It was essential, for those who sought to link sensibility to the human potential for moral behaviour, to claim that it was a universal feature of human nature. If that could not be demonstrated, then the capacity of sensibility to bear weight in moral philosophy was severely reduced. Hutcheson, in particular, devoted a huge portion of his writing to arguing that the diverse customs and moral opinions of peoples across the globe, testified by prurient travel writers, could be reconciled with a common underlying moral sense.[44] At the same time, for much of the eighteenth century, the social capital of sensibility was derived, in part, from the perception that it was a rare and precious commodity.[45] This tension between a commitment to universal sensibility and an impulse to chart its diversity and arrange it in hierarchies can be seen played out throughout the century—sometimes in the work of individual writers.

For some period thinkers, the diverse distribution of sentimental capacity was sufficient to render useless arguments that sympathy or compassion could serve as

---

[40] For discussion of these antinomies and the ways they played out in period gender politics, see Jordanova 1980; Tomaselli 1985.

[41] More 1782/1785, 285. *The Man of Feeling* was the title of a novel published by Henry McKenzie, 'the Scottish Addison', in 1771.

[42] Entry on 'Distress', *Encyclopaedia Britannica* 1798, 60–61.

[43] Smith 1759/2002, 30.

[44] Carey 1997, 276–290.

[45] Starr 1984, 126–135.

the basis for a system of social ethics. The English dissenting minister Richard Price was so disturbed by what he perceived to be the implications of Hutcheson's system, that it inspired him to publish, in 1758, *A Review of the Principle Questions in Morals*, designed to debunk the theory that moral judgements were ultimately rooted in sentiment. For Price, the key issue at stake was 'whether *right* and *wrong* are real qualities of *actions* or only qualities of our *minds*'.[46] According to him:

> If the former is true, then is morality equally interchangeable with all truth: If, on the contrary, the latter is true, then it is that only which, according to the different constitutions of *senses* of beings, it *appears* to be to them.[47]

As far as Price was concerned, by making morality 'an affair of taste', Hutcheson and Hume had rendered it entirely arbitrary.[48] For Price, moral philosophy must be grounded on something more universal. For him, that was reason or what he called 'the understanding'. Mary Wollstonecraft, another thinker influenced by the British culture of rational dissent, and one acutely aware of the way in which sensibility had been linked to period definitions of femininity, was similarly dismissive. She described sensibility as nothing but 'the most exquisitely polished instinct', insisting that 'intellect dwells not there'.[49] Others complained of the way the morality of sentiment seemed to depend on an idealised construction of human nature. The French materialist philosopher Paul-Henri-Thiry, the Baron d'Holbach, in many ways a very different thinker from Price or Wollstonecraft, remarked in his infamously atheistic *Systeme de la Nature* (1770):

> Compassion depends on physical sensibility, which is not the same in all men; it is therefore an error to make of compassion the source of our ideas of morality and of the sentiments we feel for others. Not only are men unequal in sensibility, but there are many in whom sensibility has not been developed.[50]

Where Smith and the *Encyclopaedia Britannica* feared this lack in the ignorant and brutalised masses, for the more radical d'Holbach, the absence of sensibility was most striking in 'princes, the rich, the great etc.'.[51] In his view, the alternative path towards a moral order was to teach all men that the law of nature dictated their interdependence and hence that moderation and equity would ultimately serve both their own interests and those of society at large.

For thinkers who sought to defend the universality of sympathetic sensibility despite the appearance of diversity, one way of doing so was to portray its development as a historical process built upon a natural foundation. By the second half of the eighteenth century, in particular, many had come to believe that sensibility grew in direct relationship with the process of modernisation. Despite their emphasis on the natural capacity for sympathy, many of the thinkers of the 'moral sense' school

---

[46] Price 1758/1787, 12.
[47] Price 1758/1787, 12.
[48] Price 1758/1787, 11.
[49] Wollstonecraft 1792/1994, 133.
[50] d'Holbach 1770/1774, Vol. 1, 138.
[51] d'Holbach 1770/1774, Vol. 1, 138.

were of this kind. Indeed, the association of sensibility with a civilising process can be seen even early in the century. For Shaftesbury, such sensibility was linked to the growth of politeness in the 'amicable collision' of social intercourse.[52] For Hume, Smith, and the conjectural historians of the Scottish Enlightenment, the rise of commercial society, the increase in leisure associated with the division of labour, the development of the fine arts, all nurtured sensibility and its associated forms of benevolent virtue. Indeed, for many of these thinkers, the growth of the latter amounted to nothing less than the growth of what Hume labelled 'humanity'. As he famously put it in 1752, 'refinement in the gratification of the senses' was 'the Spirit of the Age'.[53] He explained the sociology of this process:

> [People] flock into cities; love to receive and communicate knowledge; to show their wit or their breeding; their taste in conversation or living, in clothes or furniture [...] both sexes meet in an easy and sociable manner; and the tempers of men, as well as their behaviour, refine apace [...] they must feel an increase in humanity, from the very habit of conversing together.[54]

The consequence, according to Adam Smith, was that:

> Among civilised nations, the virtues which are founded upon humanity are more cultivated than those which are founded upon self-denial and the command of the passions. Among rude and barbarous nations it is quite otherwise, the virtues of self-denial are more cultivated than those of humanity.[55]

He argued that the harsh life of subsistence to which many peoples were condemned was such as to create a natural focus upon emotional and behavioural restraint in the interests of group survival. Smith was not without regret at the decline of the 'virtues of self-denial', which he saw as accompanying the growth of commercial society. He saw in it a link to the growing power of physical sensibility associated with the rise of luxury. Overall, however, he felt more than compensated by the growth of these 'virtues founded upon humanity' anchored in the refinement of that sensibility.

If the growth of 'humanity', in the sense of sympathetic sensibility, was in some sense linked to the macro processes of history for these thinkers, it did not follow that it was incapable of strategic modification. As Shaftesbury put it, 'if a natural *good* Taste be not already form'd in us; why shou'd not we endeavour to form it, and become *natural*?'[56] For Hume, despite the natural propensity of human beings for sentimental response, it was often necessary to 'employ much reasoning, in order to feel the proper sentiment'.[57] Indeed, he claimed that 'a false relish may frequently be corrected by argument and reflection' and this applied in

---

[52] Shaftesbury 1711/1999, 31.
[53] Hume 1754/1998, 167, 169.
[54] Hume 1754/1998, 169.
[55] Smith 1759/2002, 239.
[56] Shaftesbury 1711/1999, 339.
[57] Hume 1777/1975, 272–273. The *Enquiry concerning the Principles of Morals*, from which these quotes are taken, was first published in 1751.

particular to the taste for 'moral beauty'.⁵⁸ As a consequence, Hume claimed that 'the end of all moral speculations is to teach us our duty and, by proper representations of the deformity of vice and beauty of virtue, beget correspondent habits, and engage us to avoid the one and embrace the other'.⁵⁹ We can see here a clear sense that, however much sensibility might owe to the gifts of nature or the macroprocesses of history, it was also open to more wilful forms of cultivation. This widespread view promoted the emergence of what I have suggested we label affective pedagogy—a series of campaigns to cultivate the capacity for 'proper' feeling against the dangers of its alternatives.

This pedagogy could take various forms. It was pursued through everything from hygienic regimes to popular novels. The former were championed by figures such as Cheyne in England, or, on the continent, by the eminent and influential Swiss physician Samuel-Auguste Tissot. The French physician Antoine Le Camus made the physical nurturing of sensibility a core component of his *Médecine de l'Esprit* (1753). For him, it was above all the tension of nerve fibres that must be protected. This was a challenge because:

> Children, women, people who live in a rainy climate or in proximity to rivers and swamps, who live a sedentary and lazy life, who feed themselves on fatty or watery food, who have a cold and humid temperament, have naturally soft and relaxed fibres.⁶⁰

The remedy, he suggested, was to avoid all the causes which might produce these effects. While he did not specify how one might avoid childhood or femininity, he advocated living in warm, dry air, eating a dry diet, taking frequent exercise, and living in ventilated apartments as a means of optimising the sensible economy.⁶¹ For Le Camus's colleague at the Paris Faculty of Medicine, Charles Augustin Vandermonde, the cultivation of sensibility combined a species of mental training of the young, designed to nurture the quality and prevent its pathological degradation, with a program of bodily training, and partner selection for prospective parents designed to maximise the natural potential of their offspring.⁶² Unusually for the period, Vandermonde maintained there was little natural difference in the sentimental character of the two sexes.⁶³ Both, however, needed assistance to maximise their sentimental capacity from the period before conception until they reached maturity.

The literary strategy for cultivating moral sensibility was pursued by many, though in the eyes of anxious contemporaries, not all, of the novelists of sentiment who graced the period from the middle decades of the century. Samuel Richardson was widely considered the master exponent of the genre. His novels *Clarissa* and *Pamela*, in particular, were widely discussed for their affective techniques. Dr. Johnson

---

⁵⁸ Hume 1777/1975, 272–273.
⁵⁹ Hume 1777/1975, 272–273.
⁶⁰ Le Camus 1769, 223.
⁶¹ Le Camus 1769, 223.
⁶² Vandermonde 1756, vii–viii.
⁶³ Vandermonde 1756, 286. Discussed in Vila 1998, 90.

famously claimed that Richardson 'taught the passions to move at the command of virtue'.[64] Across the channel, too, Denis Diderot devoted a surprising amount of time and effort to detailing the positive moral effects of reading Richardson on both himself and his friends.[65] Rousseau, who will be discussed below, was another of the period writers to try his hand at affective pedagogy through fiction. He too, was considered a master of the art, although the moral consequences of reading his work were the subject of period dispute.[66]

For almost all advocates of sentimental training, the cultivation of sensibility required a combination of controlled exposure to experiences and protection against extreme sensations. The challenges of training sentiment were, in fact, considerable. It was not necessarily a matter of ongoing exposure to appropriate stimuli—whether it be examples of virtue, of vice, or of aesthetic beauty. As Alexander Gerard noted in his *Essay on Taste*:

> the effect of *habit* on our *perceptions* is the very reverse of that which it produces on our *active* powers. It *Strengthens* the latter, but gradually *diminishes* the vivacity of the former.[67]

The consequence of over-training, he suggested, could be 'the gradual decay of sensibility by repetition'.[68] If Gerard was worried about a loss in the capacity to feel associated with over-stimulation, many in this period were more concerned about the reverse. They feared excessive inflammation of the sensitive capacity consequent upon the dynamics of contemporary life.

For many who accepted the tie between modernity and sensibility, that tie was not purely a positive thing. If sensibility was intimately linked to the benevolent virtues, it could also be associated with a range of social ills. The growth of sedentary occupations, the shielding from labour and nature, the constant consumption of stimulating food and drink, all could lead to a kind of flabbiness of the nervous system that resulted in an excess of corporeal sensibility. The consequences could be physical weakness, nervous disorders, and even a species of solipsism that ran directly counter to the sympathetic engagement that was seen as the finest potential of sensibility in its healthy forms. Anne Vila has shown in detail how those anxieties played out in enlightenment medical philosophy and period literature in France (though they were far from confined to France).[69] I do not want to rehearse that story too much here. What I do want to emphasise, however, is the way in which for many concerned about the pathologies of sensibility, the solution came through nurturing its more sophisticated forms.

---

[64] Johnson in *The Rambler*, 97, quoted in the introduction to Blewett 2001, 7.

[65] Diderot, Denis. 1761. *Éloge de Richardson*. See also discussions in Vila 1998, 155–162; Goldberg 1984, 128–145.

[66] The issues raised were anticipated by Rousseau himself in a long dialogue written to introduce the second volume of his novel *Julie, ou la Nouvelle Héloise* (1761).

[67] Gerard 1759/1764, 100.

[68] Gerard 1759/1764, 102.

[69] The tension between the negative and positive potentials of sensibility in the eighteenth century provides both the title and the core of her book *Enlightenment and Pathology*, Vila 1998.

We can see this clearly, I think, in the case of Rousseau. He believed that there were two kinds of sensibility that were quite distinct. As he put it:

> There is a physical and organic sensibility, which, purely passive, appears to have as its purpose nothing but the conservation of our bodies, and that of our species, by the directions of pleasure and pain. There is another sensibility, which I call active and moral, which is nothing else but the faculty of attaching our affections to beings which are strangers to us.[70]

In relation to the latter, Rousseau suggested, 'the study of nerve pairs gives no knowledge'.[71] Not everyone in the eighteenth century accepted this. There were many who tried to develop physiological accounts of sympathy. Attempts can be seen in the work of the Montpellier vitalists, from whom Rousseau drew a great deal, and in the *medecins-philosophes* of the Revolut onary era. It can be seen, too, in the work of mesmerists in the last decades of the century. What was animal magnetism, after all, but a theory of organic sympathy?[72] My interest here, however, is in Rousseau's contrasting of the two kinds of sensibility. Given, as he was, to detailed self-analysis, Rousseau described the operation of the two in himself:

> Jean-Jacques seems to me to have been granted a fairly high degree of physical sensibility. It depends a great deal on his senses, and he would depend on them even more if moral sensibility did not often divert him [...] It is the mixture in the majority of his sensations that tempers them, and stealing from those which are purely material the seductive attraction of the others, makes them all act on him more moderately.[73]

The cure for excess corporeal sensibility, then, was a refinement of moral sensibility. The two were not always in competition. Indeed, moral sensibility could add to the effects of physical sensibility. In a causal reversal of the premises of Lockean epistemology, Rousseau suggested that a beautiful vista or a kindly regard did not act strongly on the external senses until they had, in some sense, pierced the heart.[74] But if physical and moral sensibility were not incommensurable, for Rousseau, it was nonetheless the case that the two existed in an agonistic relationship. He explored that relationship at length in that archetypal novel of sentiment, *Julie, ou La Nouvelle Héloïse* (1761). The passionate attraction that draws Julie and Saint-Preux together in the opening volume of that novel was, in Rousseau's mind, certainly more than physical sensibility. It was a drawing together of hearts that he linked elsewhere to moral sensibility, though he suggested it was analogous to the forces of magnetic attraction.[75] But the process by which the two lovers learn, through the second half of the novel, to set aside their passions and embrace their duties for the sake of family, friends, and the good of the community around them can be seen as a process of affective pedagogy played out within the novel. It amounts to a training of sentiment to

---

[70] Rousseau 1782, 204–205.

[71] Rousseau 1782, 205.

[72] Darnton 1968, 126–135. On broader links between ideas of magnetism and sympathy, see Fara 1996.

[73] Rousseau 1782, 207.

[74] Rousseau 1782, 207.

[75] Rousseau 1782, 205.

extend beyond its narrow focus on the object of passion. It leads not just to a rational acceptance of duty, but to a gradual recognition of the emotional impact of other beings upon the lovers' internal well-being.

We can see this interplay of competing sensibilities in other writers of the age. For many, the tutelage of sensibility was linked to a refinement of choice in the sources of pleasure. But it was by no means universal to valorise more complex forms of sensibility over simpler ones. Rousseau himself was apt to be suspicious of what he perceived as the artificial sensibilities of the *beau monde*. He saw no evidence of the connection that Hume and Smith saw between commercial society and human sympathy. From his earliest *Discours sur les arts et sciences* (1750), he had associated social refinement with a de-naturing of sentiment. And Rousseau was far from alone. His one-time friend, Denis Diderot, liked frequently to draw attention to the artificial restraints upon harmless physical pleasures he saw emerging from Catholic asceticism.[76] Jean-Paul Marat, in his pre-Revolutionary incarnation as a physician and aspiring philosopher in London, drew a distinction between physical sensibility and imaginary sensibility in his *Essay on the Human Soul* (1772). In his view, imaginary sensibility led to the pursuit of 'fictitious' pleasures, such as the love of glory or admiration, producing the excesses of war, the torment of jilted lovers. Its effect was to silence 'the voice of nature itself'.[77]

Sometimes the evaluation of different kinds of sensibility was suspended, in preference for watching their interplay in action. In Sterne's *Sentimental Journey*, for example, there is a frequent suggestion of (often barely conscious) physiological response to the environment—commonly it is of a latent erotic potential in the narrator's encounters with women. Yet that response is counter-pointed with a different kind of sensibility, a solicitous sympathy that produces a determined care for the rectitude of his relations and a desire not to distress others. As Yorrick remarks in *A Sentimental Journey*, having passed through 'The Temptation' of a pretty chamber maid in his room, and having made 'The Conquest' of his internal inclinations to take advantage of the situation:

> If nature so wove the web of kindness that some threads of love and desire are entangled with the piece, must the whole web be rent in drawing them out?—Whip me such stoics, great governor of nature! said I to myself [...] whatever is my situation, let me feel the movements which rise out of it, and which belong to me as a man, and if I govern them as a good one, I will trust the issues to thy justice—for thou hast made us, and not we ourselves.[78]

Here, once again, it is not duty or reason that does the regulating work, but feeling. The reader is rendered witness to a battle between one kind of inclination and

---

[76] Diderot's critique of the disordering metabolic effects of Catholic asceticism, and specifically monastic life, is best seen in his novel *La Religieuse* (composed about 1760, published posthumously in 1796). His views on the unnatural and physically debilitating character of European social mores can also be seen in the *Supplément au voyage de Bougainville* (composed 1772, also published in 1796).

[77] Marat 1772, 42–44.

[78] Sterne 1768/2006, 131.

another, between one kind of sensibility and another. In Sterne, who lacked the moral earnestness of a Rousseau or a Marat, the interplay between these two is often comical. The pleasures of corporeal sensibility are rather benign, and the overlay of social delicacy comes across as a charming un-worldliness. But for Sterne, as for Rousseau, the two kinds of sensibility are mixed in experience, and the restraining work of one form of sensibility upon another illustrates the same pattern.

In many ways, this gradual proliferation of the perceived modes of affect is the most striking feature of the history of sensibility over the course of the eighteenth century. While many attempts were made to theorise sensibility as a simple system producing multiple effects, there was always a centrifugal tendency within the discourse. This may well be a consequence of the fact that, for those for whom the role of disembodied reason in human life seemed increasingly limited, it was necessary to make the concept of sensibility do ever more work—at both the intellectual level and the social one. Until the very end of the eighteenth century, the culture of sensibility remained contested ground. If the philosophers of moral sense and their heirs saw in compassionate sensibility a natural inoculation against the Hobbesian world of calculation and egoism, there were many who saw in the valorisation of sentiment something that could lead to a complete breakdown of conventional order. There is, after all, only a fine distinction between the assertion that *we feel what is right* and the assertion that *what we feel is right*. Anxiety that the culture of sensibility was leading towards a dangerous belief in the latter emerged strongly in both Britain and France over the second half of the century. A critique of the personal and social costs of an excessive valorisation of the passions can be seen played out in much period literature, not least in Jane Austen's novel *Sense and Sensibility* (1811). Given these anxieties, the inevitable consequence of the rising power of sensibility was a series of campaigns to help modern men and women to learn to feel in more constructive ways. If they must inevitably be driven by their feelings, they must learn to draw from the internal sea of sensations and sentiments those fine and productive affects that would be beneficial for themselves and for society at large.

References

Barker-Benfield, Graham J. 1992. *The culture of sensibility: Sex and society in eighteenth-century Britain*. Chicago: University of Chicago Press.
Berg, Maxine, and Elizabeth Eger (eds.). 2003. *Luxury in the eighteenth century: Debates, desires and delectable goods*. Houndmills: Palgrave.
Blewett, David (ed.). 2001. *Passion and virtue: Essays on the novels of Samuel Richardson*. Toronto: University of Toronto.
Bordeu, Théophile de. 1751. *Recherches anatomiques sur la position des glandes et sur leur action*. Paris: G. F. Quillau.
Brewer, J., and Roy Porter (eds.). 1993. *Consumption and the world of goods*. London: Routledge.
Campbell, Colin. 1983. Romanticism and the consumer ethic: Intimations of a Weber-style thesis. *Sociological Analysis* 44: 279–296.
Carey, Daniel. 1997. Method, moral sense, and the problem of diversity: Frances Hutcheson and the Scottish enlightenment. *British Journal of the History of Philosophy* 5: 275–296.
Cheyne, George. 1733/1991. *The English malady*. London: Routledge.

Cook, Alexander. 2009. The politics of pleasure talk in eighteenth-century Europe. *Sexualities* 12: 451–466.
Crane, Ronald Salmon. 1934. Suggestions towards a genealogy of the 'Man of Feeling'. *English Literary History* 1: 205–230.
d'Holbach, Paul-Henri-Thiry, Baron. 1770/1774. *Systeme de la nature*, 2 vols. London.
Darnton, Robert. 1968. *Mesmerism and the end of the enlightenment in France*. Cambridge, MA: Harvard University Press.
Darwall, Stephen. 1995. *British moralists and the internal 'Ought', 1640–1740*. Cambridge: Cambridge University Press.
de Bruyn, Frans. 1981. Latitudinarianism and its importance as a precursor of sensibility. *Journal of English and Germanic Philology* 80: 349–368.
Encyclopaedia Britannica. 1798. Vol. 6. Philadelphia: Thomas Dobson.
Fara, Patricia. 1996. *Sympathetic attractions: Magnetic practices, beliefs and symbolism in eighteenth-century England*. Princeton: Princeton University Press.
Fiering, Norman S. 1976. Irresistible compassion: An aspect of eighteenth-century sympathy and humanitarianism. *Journal of the History of Ideas* 37: 195–218.
Gerard, Alexander. 1759/1764. *An essay on taste*. Edinburgh: A. Millar.
Goldberg, Rita. 1984. *Sex and enlightenment: Women in Richardson and Diderot*. Cambridge: Cambridge University Press.
Greene, Donald. 1977. Latitudinarianism and sensibility: The genealogy of the 'Man of Feeling' reconsidered. *Modern Philology* 75: 159–183.
Haigh, Elizabeth. 1976. Vitalism, the soul, and sensibility: The physiology of Theophile Bordeu. *Journal of the History of Medicine and Allied Sciences* 31: 30–41.
Haller, Albrecht Von. 1756–1760. *Mémoires sur la nature sensible et irritable des parties du corps animal*, 4 vols. Lausanne: Bousquet.
Helvétius, Claude-Adrien. 1770/1989. *De L'Homme*, 2 vols. Paris: Fayard.
Hirschman, Albert O. 1977. *The passions and the interests: Political arguments for capitalism before its triumph*. Princeton: Princeton University Press.
Hobbes, Thomas. 1651/1985. *Leviathan*. London: Penguin.
Hume, David. 1739–1740/2006. Treatise of human nature. In *Moral Philosophy*, ed. Geoffrey Sayre-McCord, 12–184. Indianapolis: Hackett.
Hume, David. 1754/1998. Of refinement in the arts. In *Selected Essays*, eds. Stephen Copley and Andrew Edgar Oxford: Oxford University Press.
Hume, David. 1777/1975. *Enquiries concerning human understanding and concerning the principles of morals*. Oxford: Clarendon.
Hutcheson, Francis. 1726. *An inquiry into the original of our ideas of beauty and virtue*, 2nd ed. London: J. Darby.
Hutcheson, Francis. 1728. *An essay on the nature and conduct of the passions and affections; with illustrations of the moral sense*. London: S. Powell & P. Compton.
Johnson, Samuel. 1755/1768. *A dictionary of the English language*. Dublin: Thomas Ewing.
Jordanova, Ludmilla. 1980. Natural facts: A historical perspective on science and sexuality. In *Nature, culture and gender*, ed. Carol MacCormack and Marilyn Strathern, 42–69. Cambridge: Cambridge University Press.
Kaiser, Thomas. 2010. *Enlightened pleasures: Eighteenth-century France and the New Epicureanism*. New Haven: Yale University Press.
Lacratelle, Charles (ed.). 1789. *Encyclopédie méthodique, ou par ordre de matiéres: Logique, metaphysique et morale*, vol. 3. Paris: Panckoucke.
Langford, Paul. 1989. *A polite and commercial people: England 1727–1783*. Oxford: Oxford University Press.
Le Camus, Antoine. 1769. *Médecine de l'esprit*, 2nd ed. Paris: Ganeau.
Mandeville, Bernard. 1714/1989. *The fable of the bees: Or private vices, public benefits*. Harmondsworth: Penguin.
Marat, Jean-Paul. 1772. *An essay on the human soul*. London: T. Becket & Co.
Montesquieu, Claude-Louis-Secondat, Baron de. 1748/1768. *De L'Esprit des lois*, 4 vols. London.

More, Hannah. 1782/1785. *Sacred dramas: Chiefly intended for young persons ... to which is added, sensibility, a poem*. Dublin: P. Byrne.

Mornet, Daniel. 1929/1969. *French Thought in the Eighteenth Century*. Trans. Lawrence M. Levin. New York: Archon.

Nederman, Cary. 1988. Nature, sin and the origins of society: The Ciceronian tradition in medieval political thought. *Journal of the History of Ideas* 49: 3–26.

Norton, Robert. 1995. *The beautiful soul: Aesthetic morality in the eighteenth century*. Ithaca: Cornell University Press.

Pocock, John. 1975. *The Machiavellian moment: Florentine political thought and the Atlantic republican tradition*. Princeton: Princeton University Press.

Price, Richard. 1758/1787. *A review of the principle questions in morals*, rev. ed. London: T. Cadell.

Radcliffe, Elizabeth. 2004. Love and benevolence in Hutcheson and Hume's theory of the passions. *British Journal of Philosophy* 12: 631–653.

Rousseau, Jean-Jacques. 1782. Dialogue de Rousseau, juge de Jean-Jacques. In *Collection complète des oeuvres de J. J. Rousseau*, vol. 11. Geneva.

Shaftesbury, Anthony Ashley Cooper, Earl of. 1711/1999. *Characteristics of men, manners, opinions, times*. Cambridge: Cambridge University Press.

Sheridan, Thomas. 1797. *A complete dictionary of the English language*, 4th ed. London: Charles Dilly.

Skinner, Quentin. 1978. *Foundations of modern political thought*, 2 vols. Cambridge: Cambridge University Press.

Smith, Adam. 1759/2002. *The theory of moral sentiments*. Cambridge: Cambridge University Press.

Starr, George A. 1984. Egalitarian and elitist implications of sensibility. *L'Égalité* 9: 126–135.

Steinbrügge, Lieselotte. 1995. *The moral sex: Woman's nature in the French enlightenment*. Oxford: Oxford University Press.

Sterne, Lawrence. 1768/2006. *A sentimental journey through France and Italy*. Indianapolis: Hackett.

Tomaselli, Sylvana. 1985. The enlightenment debate on woman. *History Workshop Journal* 20: 101–124.

Vandermonde, Charles Augustin. 1756. *Essai sur la manière de perfectionner l'espèce humaine*. Paris: Vincent.

Vila, Anne C. 1998. *Enlightenment and pathology. Sensibility in the literature and medicine of eighteenth-century France*. Baltimore: Johns Hopkins University Press.

Voitle, Robert B. 1955. Shaftesbury's moral sense. *Studies in Philology* 2: 17–38.

Wollstonecraft, Mary. 1792/1994. A vindication of the rights of woman. In *A vindication of the rights of woman & a vindication of the rights of man*, 63–284. Oxford: Oxford University Press.

Chapter 6
Physician, Heal Thyself! Emotions and the Health of the Learned in Samuel Auguste André David Tissot (1728–1797) and Gerard Nicolaas Heerkens (1726–1801)

Yasmin Haskell

Abstract The Dutch physician and Latin poet, Gerard Nicolaas Heerkens (1726–1801), published in Groningen in 1790 an expanded edition of his Latin didactic poem on 'the health of men of letters' (*De valetudine literatorum*), which he originally composed as a medical student in Paris some 40 years earlier and published in 1749. Heerkens's work belongs to a long tradition of humanist theorising about the occupational health of the learned. In the years between the first and second editions, Samuel Auguste André David Tissot (1728–1797), Lausanne physician, professor, and public health advocate, had also published a Latin academic oration on 'the health of men of letters'. Heerkens does not neglect to assert the priority of his own *De valetudine literatorum*. Tissot's oration stigmatised as pathological precisely the sort of life of learning in which Heerkens himself was engaged. This chapter reviews Heerkens's rather testy engagement with Tissot, his defence of the passion for learning, and his advice to the learned on moderating their passions.

The Dutch physician and Latin poet, Gerard Nicolaas Heerkens, published in Groningen in 1790 a revised edition of his Latin poem, in three books exceeding a thousand verses apiece, on 'the health of men of letters' (*De valetudine literatorum*). The poem was originally composed when he was a medical student in Paris some forty years earlier, and published in a much truncated form in Leiden in 1749.[1] Heerkens's work belongs to a long tradition of humanist theorising and worrying about the occupational safety of the learned.[2] In the years between the first and second editions, Samuel Auguste André David Tissot (1728–1797), Lausanne physician, professor, and public health

---

[1] See Haskell 2007a.
[2] See Kummel 1984; Mikkeli 1999.

Y. Haskell (✉)
School of Humanities, ARC Centre of Excellence for the History of Emotions, 1100–1800, University of Western Australia, 6009 Crawley, Perth, WA, Australia
e-mail: yasmin.haskell@uwa.edu.au

advocate, best known to posterity for his writings on migraine and masturbation, had also published a Latin academic oration on 'the health of men of letters' (*Sermo inauguralis de valetudine litteratorum habitus public die 9 Aprilis 1766 cum novam medicinae cathedram auspicaretur*). Heerkens does not neglect to assert the priority of his own *De valetudine literatorum*.[3] It must have been galling for him to see Tissot's work go through not just a second Latin edition (Lausanne, Frankfurt, and Leipzig, 1769) but no less than five in French, the first unauthorised, by the date of the second edition of his poem; and all the more so, since Tissot's oration seemed to denigrate the very life of learning in which Heerkens himself was engaged.[4]

In the second book of his poem, having warned overambitious parents against hot-housing their young children, Heerkens is careful to balance his speciously Rousseauian strictures with a respectful nod to the dignity of learning:

> You who help to ward off disease [sc. physicians] should not, to be sure, mock and entertain yourself at the expense of the chorus of scholars. You act disgracefully if you relay any precepts and prescriptions for them in a sarcastic manner. Let that—since he blathered laughably in Latin, and that's why he cannot bear scrutiny by critics and philologists—let that be considered a badge of honour by Tissot alone, he who has been laughed at more than once for his remedies. He paints the whole of Pindus for you, and the court of Pallas, as brimming with the sick and the mad.[5]

The literature on the occupational health of scholars was extensive by the middle of the eighteenth century.[6] In the first edition of his poem, Heerkens cites, as an *envoi*, Marsilio Ficino, Vobiscus Fortunatus Plemp, and Bernardino Ramazzini as reliable authorities for further reading.[7] In the second, he again recommends Ramazzini (1633–1714) for his work on the diseases of craftsmen, monks, and the health of princes; also Jodocus Lommius (c. 1500–c. 1564) for his commentaries on the first book of Celsus on preserving health (Leuven, 1558); Guglielmo Gratarolo (?1516–?1568) for his 'On preservation of the memory' (Zurich, 1554) and 'On conserving and preserving the health of men of letters and those in public office' (*De literatorum & eorum qui magistratibus funguntur conservanda praeservandaque valetudine*, Basel, 1555); Santorio Santorio (1561–1636); and Francis Bacon (1561–1626), especially his 'History of life and death' (*Historia vitae et mortis*, 1623).[8]

---

[3] 'Tissot, who published his little work "On the health of men of letters" seventeen years after mine' (*Tissotius, qui xvii post meum annis, suum de Literatorum Valetudine opusculum vulgavit*) Heerkens 1790, 105, N. 68.

[4] By this date, of course, there was a very extensive literature in German-speaking lands about the habits, passions, and vices of the learned. See Hummel 2002.

[5] '*Non certe a morbis decet irridere juvantem/Aut cupidum studio te recreare chorum./Turpiter id facis, si qua ulla, facetus inepte,/Praecepta huic dictas, auxiliumque feres./Id, quia ridiculus Latio blateraverat ore,/Nec criticos ideo, grammaticosque ferat,/Ducat id esse sibi Tissotius unus honori/ Risus ab auxiliis non semel ipse suis./Ille omnem pingit tibi Pindum, et Palladis aulam,/Ut plenam aegrotis, mente inopemque domum.*' Heerkens 1790, 60. All translations are my own.

[6] The classic departure point of this tradition is Marsilio Ficino's 'De studiosorum sanitate tuenda' (the first book of his *De vita libri tres*). Ficino 1489.

[7] Heerkens 1749, 18.

[8] Heerkens 1790, 127.

There was also a venerable tradition of Latin *poetry* by physicians on the theme of medical 'hygiene', or preservation of health.⁹ As poetic precursors in his chosen genre, Heerkens cites, in the first edition, Malcolm Flemyng's Lucretian didactic *Neuropathia*, on hypochondriac and hysteric disease (York, 1740) and his own medical professor Adriaan van Royen's elegiac 'loves of the plants' (Leiden, 1732).¹⁰ He also coyly alludes to a didactic poem by French physician, Claude Quillet, on the begetting of beautiful children: 'the Pierian muses do not do well to reveal the secrets of Cythera; Thalia would be tough on a new Quillet'.¹¹ None of these works is really exploited as a literary or scientific model by Heerkens, however. He is an admirer of Ovid, and in the shorter and sweeter first edition of his poem we get at least a taste of the latter's wit. In the section advising sexual continence for scholars, for example, Heerkens suggests that once they have run out of saints' days, they should observe poets' birthdays, such as Ovid's (!), as a pretext for excusing themselves from their conjugal duties.

The topics treated in the telescopic 1749 edition of the *De valetudine litteratorum* are traditional, or at least uncontroversial.¹² We might note, however, the relative *absence* of discussion of 'melancholic', 'hypochondriac', or 'nervous' illness, the disease(s) of scholars par excellence in the early modern period.¹³ The restricted compass of the first edition (only 366 verses) precluded any detailed dietary or pharmacological advice such as was to be found in hygiene writers from the Renaissance through the eighteenth century. But it transpires that the second edition, some ten times longer, is also relatively nontechnical in this respect. While the poet-doctor availed himself, there, of all the space he could ever have wished for, the result is no versified textbook or prescription pad.¹⁴ Rather, the 1790 edition is crammed with

---

⁹ See Fischer 1988a, b; Haskell 2008.

¹⁰ Van Royen's poem 'on the diseases of the ages' seems to have been first published in 1771.

¹¹ '*Non bene pierides vulgant arcana Cytherae:/Quilletio novo dura Thalia foret*'. Heerkens 1749, 12. On Quillet, see Ford 1999. When he condemns the abuse of tea and Pierre-Daniel Huët's poetic praise of it, Heerkens does not here cite a well-known little didactic poem on tea by French physician and Latin Pleïade member, Pierre Petit, *Thea ... carmen* (Paris, 1685)—a surprise, as the author is mentioned in Heerkens's introduction as a medical authority (Heerkens 1749, 8–9).

¹² See Haskell 2007a.

¹³ In fact, the same is largely true for the second edition. There Heerkens simply observes *en passant* that lack of sleep leads to stagnation of food in the stomach, enfeeblement of the body, and embittering of the blood, and that 'there is no other lamentable cause for the melancholy disease and, like the blackish bile, consumption itself proceeds from this [sc. the bitter blood]'. ('*Causa melancholico non flebilis altera morbi,/Bilis et ut nigricans, hinc phthisis ipsa venit,*' Heerkens 1790, 121.) Perhaps his most interesting observation on 'hypochondria' is relegated to a footnote: that 'valetudinarians, melancholics, and hypochondriacs' who are confined to bed and a frugal diet are observed to live long lives (Heerkens 1790, 167–168, N. 39)!

¹⁴ For example, in the context of a discussion on raising children in the first book, Heerkens avails himself of the didactic-poet's prerogative of 'passing over' (*praeteritio*): 'I will not touch on diet: it is prescribed by many—and the kind that's harmful, the Fate that snatches away so many boys'. ('*Victum ego non tango: multis praescribitur: isque,/Quod noceat, pueros Parca tot usque rapit*', Heerkens 1790, 14.) Specific dietary advice is held over until the final pages of the third book, almost as an afterthought.

anecdotes and digressions, for example on Heerkens's education, his personal encounters with *philosophes* in Paris and observations on their characters, his appraisal of various Dutch poets, scholars, and painters, Dutch university reforms, local archaeology, and even on the fortunes of the Groningen beer industry.

In this chapter, I will compare the respective emphases of Heerkens's 1790 poem and the enlarged French edition of Tissot's oration published in Lausanne in 1769 (to which Heerkens refers).[15] Tissot's *De la santé des gens de lettres* which, after all, had its birth as an academic oration, on the occasion of the author's assuming the Chair of medicine at Lausanne, opens with a grand preamble on the nobility of medicine, establishing its links with other provinces of learning from divinity to law to natural philosophy to languages and history. Tissot proceeds to identify two sources of illness in the learned: overuse of mind and under-use of body.[16] Within a few pages, the copious professor is leading us through a rogues' gallery of mostly anonymous scholarly invalids drawn from the collected case histories of medical colleagues past and present: Van Swieten on the man who became dizzy if he listened too attentively to a story, even a frivolous one, and was exceedingly anxious when trying to recall to mind everything he had forgotten; Bordeux on one whose arm swelled up whenever he was thinking or experiencing an intense emotion; Pechlin on the woman who had a little fit when she concentrated too hard on reading or writing; Morgagni on the scholar who got a nosebleed from meditating on abstract matters before rising, and on the preaching monk who died of apoplexy mid-sermon.[17] Tissot's friend, Zimmerman, reports on 'the too interesting literary exhaustion' (*l'épuisement littéraire trop intéressant*) of a young Swiss gentleman whose musings on metaphysics induced a complete stupor lasting a year,[18] and on a Swiss pastor who, eager to maintain the high reputation he had gained for his sermons, through excessive reading and attention to their composition gradually lost his memory.[19] Tissot claims to have seen sick scholars who, through their 'literary intemperance' had lost all appetite for food so that their digestive functions had ceased; they had become feeble, wracked by convulsions, and had finally lost all their senses.[20] While Heerkens is by no means blind to the dangers of excessive study, especially when combined with lack of exercise and bad diet, there is nothing in his poem to match this undignified catalogue of learned woe.[21]

---

[15] Anne C. Vila conducts a parallel calibration of Tissot's treatise with those of two contemporary rivals, the Paris university physicians, Le Camus (*La Médicine de l'esprit*, query 1753; definitive edition 1769) and Vandermonde (*Essai sur la manière de perfectionner l'espèce humaine*, 1756). Vila 1998, 80–107.

[16] Tissot 1769, 14.

[17] Tissot 1769, 21, 22, 34, 46.

[18] Tissot 1769, 23. 'Without being blind, he seemed not to see; without being deaf, he seemed not to hear; without being mute, he ceased to speak'. ('*Sans être aveugle, il paroissoit ne pas voir; sans être sourd, il paroissoit ne pas entendre; sans être muet, il ne parloit plus*'.)

[19] Tissot 1769, 52.

[20] Tissot 1769, 28–29.

[21] When warning against the dangers of the sedentary lifestyle, especially for residents of damp climes, Heerkens recalls, in the first book, his impressions of an enclosed community of nuns in

Indeed, Tissot's cabinet of curiosities verges on a chamber of horrors when he invites us to confront the spectre of delusional madness. He has merely to conjure the name of Torquato Tasso to cast a chill, and he elaborates the example of the painter Spinello, whose depiction of Lucifer was so terrifying that the image he had created haunted him for the rest of his life.[22] The Dutchman Caspar Barlaeus, poet, orator, and physician, knew the perils of mental exertion, and warned his friend, Constantijn Huygens, against replying to his letter, lest the effort precipitate a fresh health crisis—but Barlaeus could not save himself, became convinced that he was made of butter, and committed suicide in a well.[23] As for Tissot, he claims to have grieved first-hand for a friend whose devotion to letters and medicine led to his total derangement through unremitting study. Religious zeal had similar effects, as in the case of the woman who embraced the Moravian cult and was reduced to repeating 'my sweet lamb' over and over again, until she died.[24]

Heerkens, too, was alive to the dangers of fanatical religion. He tells us that a 'little spark' (*scintillula*) of madness planted in the developing mind can derange it, leading inexorably to the mental hospital. The poet even confesses having found himself in this dangerous position as a youth, when he was sent away to the Jesuit college at Meppen:

> The prefect of the Westphalian college had inspired me as a young boy to take for qualms of conscience, and for real and very serious sins against chastity, those things that should not be considered qualms of conscience or the shadows of sin. And since those things afflicted my mind for almost a whole year, and to such an extent that I dared not raise my eyes, when I returned to my father during the vacation and seemed not sufficiently to heed his advice, I was entrusted to a physician [Eutropius Eiding] whom I knew to be greatly respected by my father: and by that man's clear arguments, together with strong words, I was restored within a week or two to a rational perspective on my doubts, and I was sent back to school—but into the care of a wiser director of conscience.[25]

It is not just benighted Catholic priests who are to blame for such terrors. Heerkens warns 'sacred doctors' that they risk ruining the 'minds of poor little boys, and their own children. They prate of nothing but hell, both in the churches and at

---

Bruges. The uncle of a Dutch friend was their director, and the student had been commended into his hospitality when visiting the city for a few days; he was also invited to observe a young girl entering the novitiate. Heerkens was amazed to find that this spiritual director, 'although an educated man, and most learned in Hebrew, was so naïve (*tanta erat simplicitate*) as to believe that all women of a certain age went mad, and was persuaded of this by the example and spectacle of his own girls. And I pitied them all the more since male initiates of the same order [sc. the Carthusians] in Belgium were allowed to go out of doors' (Heerkens 1790, 16, N. 18).

[22] Tissot 1769, 40.

[23] Tissot 1769, 41–42. See Blok 1976. For Heerkens's account of Barlaeus, see below.

[24] Tissot 1769, 45. Tissot refers to many interesting observations on religious melancholy (*la mélancholie devote*) assembled by his friend Zimmerman, 'of which the symptoms are as bizarre, as frightening, as cruel as could be possible'. ('*dont les symptoms sont aussi bizarres, aussi effrayants, aussi cruels qu'il soit possible*', Tissot 1769, 133.)

[25] Heerkens 1790, 151, N. 23. In the Latin text, Heerkens's autobiographical anecdote is related in the third person.

home, often heavy with sleep, and after drinking'.[26] That he had Dutch Calvinists in mind here is clear from his characterisation of their view of salvation or damnation as something that 'may not be averted by any pains, nor golden virtue, nor a humble life of continuous prayer to God'.[27] Those so fixated on their posthumous fate cannot escape a degree of madness. And although he was taught 'by Rome, remote from Dordrecht',[28] that good deeds counted for something in placating the deity, Heerkens's youth and good manners had prevented him from resisting the prevailing view: 'the edict was enough: and for those denying ears to the decree, there was Jungius before their eyes, and a crowd to boot [i.e. of hard-line Protestants]'.[29]

On the other hand, Heerkens commends the health benefits of the *right* kind of religion.[30] When he was about to depart for his studies abroad, and the burgomeester of Amsterdam enquired of him what resources his father had put at his disposal, Heerkens replied that 'he advised me to imagine him, and especially God, always and at every hour, by my side'.[31] As exercise benefits the body, so does deep peace the mind. The poet asks, ruefully,

> what dues does death deliver to other good souls [sc. morally upright atheists]? An eternity of slumber, to be sure, under the dark soil! The mind consoling itself with deep sleep will look forward to this, or to that which it was before the time of its birth.[32]

Such nihilism, he observes, has little appeal at the final hour, even for philosophers. A pragmatic psychologist, Heerkens recognises the protective value of religious faith regardless of its objective truth. Conceding that some unbelievers are righteous

---

[26] '[…] *miseris ut saepe puellis,/Et mentem natis quod violentque suis./Nil prius, ac fati tenebras, templisque domique,/Saepe graves somno, cum biberintque, crepant*'. Heerkens 1790, 152.

[27] '*Idque nec avertat cura ulla, nec aurea virtus,/Continua supplex nec prece vita Deo*'. Heerkens 1790, 152.

[28] A reference to the Synod of Dort, where the traditional Dutch Calvinists confronted the Arminians.

[29] '*Sufficit edictum: dictoque negantibus aures/Jungius ante oculos, parque caterva manet*'. Heerkens 1790, 152. Joannes Ernestus Jungius, as Heerkens explains in a note, was a preacher who published in Zutphen in 1749 an anti-papal eschatological commentary, and declared himself to be a king, manifesting sure signs of election by God. After this mad work was reprinted ten years later, Jungius was removed from his ministry and confined in a mental hospital.

[30] In a satire addressed to Burgomeester of Nijmegen, Cornelius Walraven Vonck, Heerkens writes scathingly of self-mortification, of 'the ape of the mob of Perugia [who] cuts and wounds his back with scourges; and he teaches the whole congregation to weep and suffer as if gentle Religion decreed that it was a virtue for Christians to be miserable. There is no need for good people to cry, groan, and fear; a gentler and more even road leads the blessed to Heaven'. ('*Perusini simia vulgi,/Terga flagris caedit, lacerat; plenoque theatro/Flere patique docet; virtutem Christicolasque/Relligio miseros tanquam esse benigna juberet./Non opus est lachrimaque, bonis, gemituque, metuque;/Mollior ad Caelum via ducit & aequa beati!*' Heerkens 1751, 5). A note to these lines points to the first chapter of L.-J. Lévesque de Pouilly's *Theorie des sentimens agreables* (Lévesque 1747). See below.

[31] '*Consuluit, dixi, se semper, et ad latus horis/Omnibus ut videam, praecipueque Deum*'. Heerkens 1790, 153.

[32] '*Ecqua bonis aliis mors debita mentibus offert?/Saecula sub nigra nempe quietis humo?/Praedicet hoc, altum mens se solata soporem,/Aut quod natales temporis ante fuit*'. Heerkens 1790, 154.

men, he suggests that they would be much happier if only they believed: 'I believe that good men who are atheists have received just rewards for pious deeds—but as soon as the mind which is unable to trust in the providence of heaven finds itself afflicted, it is more sorely downcast'.[33]

These reflections lead to a revealing section in which Heerkens contrasts the ancient and modern philosopher. He reports that

> quite a few of the ancients thought that [to believe in] life after death was to enjoy a desirable credulity. And intelligent people [today] struggle *not* to have that belief! And the intelligent wish to snatch it away from the rest![34]

Such scepticism, he continues, may have been forgivable in the ancients, since their religious understanding was obscure, 'and yet the truth [sc. of life after death] *was* revealed at that time, and not just so that men might live more securely in the midst of so many crimes and deceptions'—Heerkens alludes here to the freethinking notion of posthumous punishment as an instrument of social control—'but as a great and philosophical truth: whence Cato drew solace for his fate at the final hour. To be sure, it is reported [sc. in Plutarch] that in ancient times many people genuflected whenever they saw a statue of Plato'.[35] How different, exclaims Heerkens, were those pious philosophers of old from our moderns! Not that he would wish the harsh fate of Vanini or Calaber Calabro on today's freethinkers; he will, however, exhort them to hold their tongues.[36] Whether or not they are personally convinced of the existence of God they should take thought for the detrimental social consequences of airing their views in public: 'I wish they would conceal the opinions suggested to them by an unbelieving mind—and they will be an unspeakable sore of the mind—and that they would conceal even those opinions they consider to be true'.[37] The phrase 'unspeakable sore of the mind' (*infandum mentis [...] ulcus*) has

---

[33] '*Credo bonos, Superis sed quorum incredula mens est,/Praemia de factis justa tulisse piis:/Sed, simul auxilio mens fidere nescia caeli,/Se videt afflictam, tristius aegra jacet*'. Heerkens 1790, 153–154.

[34] '*Et veterum haud paucis, vitam post fata morari,/Visum est optanda credulitate frui./Et bona mens quibus est, sibi desit ut illa, laborant!/Et bona mens aliis hanc rapuisse velit!*' Heerkens 1790, 155.

[35] '*Et tamen ostensum est, neque tantum ut tutius inter/Tot scelera et fraudes vivere cuique foret:/ Grande sed ut verumque sopho! Sua sorsque Catoni/Extremo fieret mitior unde die./Scilicet, ut fama est, et flexo poplite quondam/Ante oculos multis signa Platonis erant*'. Heerkens 1790, 155.

[36] In an accompanying note, though, he writes more approvingly about the punishments meted out to them—death and long imprisonment, respectively. I have been unable to glean any biographical information on 'Calaber' beyond Heerkens's indications here: that he was a foreigner whom the author had known in Rotterdam—not as an impious writer but as a mentally ill man who spoke his mind too freely about affairs in his host country; that he was 'very well known to the Dutch' because he was condemned to serve twenty years in the 'Gaudium' prison for speaking out against the state religion. Such punishment 'has been meted out more than once to writers undermining religion in the hearts of men, and rightly so, and in all countries, to deter their outspokenness'. ('*scriptores religionem in animis hominum minuentes, haud semel, et jure, et in omnibus terris secuta fuit ... ad deterrendam eorum licentiam*', Heerkens 1790, 157, N. 27).

[37] '*Opto, suos celent, si quos incredula sensus/Mens daret: infandum mentis et ulcus erunt/Et celent, veros etiam quoscunque putarint*'. Heerkens 1790, 157.

something of Voltaire's 'écrasez l'infame' about it, but it is, in fact, another modern philosopher that Heerkens singles out in this context: Spinoza.

Heerkens's argument, perhaps surprisingly, is not so much that atheists should fear for the fate of their souls: rather, the expression of their views is dangerous to their own and to public mental health. Those who spread the *virus*[38] of unbelief will come to regret it: 'I have seen men whose minds have been overturned by that pain, to whose faces it has given a wild and ashen aspect'.[39] These unfortunate philosophers will see some driven mad (*laesum caput*) by their views, many more affected in their hearts (*cor*), 'and not one of their evil disciples right in the head'.[40] This was the fate of Spinoza, 'and they say he testified to that with a bitter groan, though he had feared nothing until the point of death'.[41] A fascinating note informs us that Heerkens's contention in these verses was publicly approved by eminent Groningen Orientalist, Albert Schultens, on the authority, in turn, of an 'erudite and noble old man' familiar with Spinoza towards the end of his life. This acquaintance had declared the philosopher to be wholly different from his caricature in hostile posthumous rumour. Spinoza was in no way shifty in appearance or morals, Schultens's source reported, but was candid and blameless. Heerkens thus defends the credibility of his poetic claim of a deathbed recantation—also, because Spinoza had deplored the bad behaviour of his own disciples. However, when the young Heerkens had originally recited this note to his literary advisors in Paris:

> I remember that it was objected that Spinoza considered his disciples' bad morals as nothing more than a storm which had blown up in the universe, of which he had taught that both he and they were mere particles, and therefore that [statement of mine] that he was squeamish about the morals of this or that disciple was unbelievable. But I pointed to Pierre Bayle's assertion that Spinoza had forbidden his disciples to use his name, and that it therefore appeared that he considered the storm arising from himself to be of a different nature from that proceeding from the laws of the universe; and that he was not able to claim any right for himself to whip up cosmic storms, nor, on that account, perhaps even to wish them. But some wit quipped: 'Good weather usually follows a storm!' Not for those, I replied, whom the storm has overwhelmed. And good weather follows the storms made by *God*, not those arising from men, and from perverse philosophers.[42]

In the verse text, Heerkens runs a curiously materialistic justification for religious faith, a sort of up-dated and sophistical version of Pascal's wager: Spinoza died of disease, but his mind was not impaired by old age. Nevertheless, it is established that *both* fever and age can alter the mind. Imagine the terror we would experience if we had to confront death and eternity with our philosophical 'belief' (*opinio*) no longer in conformity with our 'changed nerves' (*mutatis [...] nervis*)! Heerkens then reprises the theme of 'do no evil'. Those who are pricked by religious

---

[38] Heerkens uses this word three times in almost as many pages. In this period, of course, it does not mean 'virus', but the sense of corrupting poison still marks it as a medical term.

[39] '*Vidi, queis mentem dolor ille everterat omnem,/Et vultum dederat canitiemque feram*'. Heerkens 1790, 157.

[40] '*unum/Nec de discipulis mente valere malis*'. Heerkens 1790, 157.

[41] '*Idque, licet sub fata nihil metuisset, amaro/Spinosam gemitu testificasse ferunt*'. Heerkens 1790, 157.

[42] Heerkens 1790, 157–158, N. 28.

doubts but keep silent may hope for divine mercy, but what mercy can be expected for the 'spreader of wicked ideas, and the pestilential philosopher whose mouth has been persuaded to evil'?[43] Better to have been a tyrant or a scourge of one's country than to have left a legacy of corrupting souls after one's death! By and large, Heerkens takes a pragmatic and 'rational' view of faith more or less consonant with that of the early Voltaire.[44]

Interestingly, Heerkens does *not* invoke the late-Renaissance language of religious melancholy when discussing religious fanaticism.[45] Tissot, even as he uses the more contemporary terminology and metaphorical economy of the 'nerves'—thus he writes of tension and slackening of fibres, of overuse of brain parts—continues to draw on a rich classical and late-Renaissance heritage of humoral medicine, and on frightening disease concepts from 'melancholy' through to 'hypochondriac' disease.[46] Moreover, Tissot claims that nervous disease has become especially prevalent in the past 60 years—but that melancholy and especially 'hypochondriac' illness was proliferating is, in fact, a common refrain going back to the late Renaissance.[47] And if he distinguishes (the newer) 'nervous' from the more familiar visceral variety of hypochondria, he does not depart radically from a long tradition of humanist hygiene which essentially saw the 'organs below the cartilage of the ribs' as the crucible of scholarly distemper.[48] Tissot reports of the great Dutch

---

[43] '*doctrinae [...] vulgator iniquae,/Et male persuaso pestifer ore sophus*'. Heerkens 1790, 159.

[44] His views fit comfortably within the framework outlined by David Sorkin. Significantly, as Sorkin notes, 'many of its fundamental ideas, Protestant and Catholic, first appeared in the Dutch Republic, which maintained a precarious toleration' (Sorkin 2008, 6).

[45] On early modern religious melancholy see Gowland 2006a; Schmidt 2007.

[46] 'Among the problems that this great quantity of humours engenders in the brain, let's not forget that it contributes not a little to that unfortunate condition which produces the hypochondriac disease; the fibres of the brain become weak from dilatation, become more soft and incapable of resisting various impressions, which gives rise to the hypochondriac character.' ('*Parmi les maux que cette grande quantité d'humeurs cause au cerveau, n'oublions pas qu'elle contribue beaucoup à cette malheureuse disposition qui produit l'affection hypocondriaque; les fibres du cerveau en se dilatant s'affoiblissent, deviennent plus molles & incapables de résister aux différentes impressions, ce qui fait le caractère de l'hypocondrie nerveuse*', Tissot 1769, 55–56.) We divide this illness into two kinds; that which is simply nervous we have seen above is the result of concentration [*contention*]; and that which depends on the distention of the abdomen and disturbance of the digestion is the regular result of lack of mobility '*On divise cette maladie [sc. hypochondria] en deux especes; celle qui est simplement nerveuse, nous avons vu plus haut qu'elle étoit l'effet de la contention; & celle qui dépend de l'engorgement des viscères du bas-ventre & du dérangement des digestions; elle est l'effet constant de l'inaction*'. (Tissot 1769, 75–76)

[47] See Gowland 2006b; Haskell 2007b, 2011b. Something of our modern sense of hypochondriac health anxiety is captured in Tissot's concern that convalescing scholars are creatures of habit and are prone to becoming obsessional about their health regimes—although he does not call *this* 'hypochondria' (Tissot 1769, 253–254).

[48] Cf. Aretaeus of Cappodocia: 'There are other, and, indeed, innumerable causes of this disease; but the principal is, much pus poured forth by the belly through the stomach [...]. It is familiar to such persons as from their necessities live on a slender and hard diet; and to those who, for the sake of education, are laborious and persevering; whose portion is the love of divine science, along with scanty food, want of sleep, and the meditation on wise sayings and doings' ('On the causes and signs of acute diseases', Aretaeus 1856/1972, Book 2, Chap. 6).

microscopist Jan Swammerdam that 'this capable observer of Nature was so tormented by melancholy, or the black bile, that he scarcely deigned to reply to those who addressed him'.[49] Shortly before his death, Swammerdam was seized by a 'melancholic fury' (*fureur mélancolique*) and burned all his writings, after which he wasted away to a skeleton.[50] It is true, concedes Tissot, that this sort of melancholy, which enables a man to remain fixed on one idea, to consider it from all angles, and without distraction, has long been observed to be useful for scholarship—but at what cost to human health and happiness?

Like Tissot, Heerkens commends society, cheerfulness, and, as we have seen, religious faith, for the modern man of letters. Superficially, our two authors cover a lot of common ground besides: they both warn against the dangers of abuse of tobacco and tea, of changes in the weather, of late-night study vigils; they both advise a frugal diet and daily exercise[51]; they both rail against parents who ruin the health of their children with unrealistic educational expectations. But Tissot sees no value in learning unrelated to a child's future occupation—and for the most part, learning is not to be cultivated as an occupation in its own right! As in various near-contemporary German writers discussed by Pascale Hummel, like Johann Andreas Fabricius,[52] one detects an almost perverse delight in Tissot's rehearsal of the curious and sordid bodily symptoms associated with abuse of the life of the mind: from bad breath, mouth ulcers, flaccid skin, and gum disease, through alopecia, gallstones, flatulence, shortness of breath, haemorrhoids from long sitting, and constipation as a result of retaining faeces in order to continue studying.[53] Tissot's scholar is a wretched and even repellent creature, a consequence of whose poor personal hygiene is obstructed perspiration,[54] whose semen is impoverished and incapable of producing illustrious sons, whose unhealthy lifestyle is as difficult to correct as that of the lover who is told that the beloved is flawed.[55] This last observation is telling: the physician Tissot writes about the 'passion' for learning as if it were, in a way, a variant of Renaissance love melancholy.[56] For Tissot there is almost always something *pathological* about learning. In the second half of his treatise (some 130 pages) he proceeds from symptoms to remedies, prescribing a careful diet, various

---

[49] '*cet habile observateur de la Nature, [...] étoit tellement tourmenté par l'atrabile ou bile noire, qu'à peine daignoit-il répondre à ceux qui lui parloient*'. Tissot 1769, 76.

[50] Tissot 1769, 77.

[51] The younger Heerkens knew from personal experience the excruciations of the scholarly stomach. During his studies he had recourse to various medicines and healing waters, to no avail. Relief came, in the end, *not* from abandoning his studies, but from an unexpected source: his long rambles around suburban Paris (Heerkens 1790, 19–20).

[52] The relevant text is Johann Andreas Fabricius. 1752–1754. *Abriss einer allgemeinemn Historie der Gelehrsamkeit*. Leipzig. See Hummel 2002.

[53] But his long meditations also, apparently, have the effect of excessive evacuation. (Tissot 1769, 38).

[54] Tissot 1769, 97. On this subject, Tissot recommends a dissertation by German physician I. Z. Platner. 1731. *de morbis ex immunditiis* (on diseases from lack of cleanliness). Leipzig.

[55] Tissot 1769, 84, 136.

[56] See, for example, Ferrand 1991.

drugs, purging, moderate bloodletting, rubbing, spas, and forbidding even short trips to libraries! An intellectual life might be appropriate for true geniuses and for those blessed with the constitution for it, but unfortunately, since the Renaissance, the world has been overrun by '*érudits*', a species of men unknown in antiquity that tortures itself like the *Fakirs* of India—but with cold as opposed to sunshine, with manuscripts, medals, and inscriptions, and physical inactivity, as opposed to whips and chains.

Moreover, most modern learning is *futile*. The majority of today's scholars, Tissot exasperatedly reminds us, are not destroying their health to produce anything so useful or interesting as Montesquieu's *Esprit des Lois*:

> One compiles all the most commonplace things; the other says again that which has been said a hundred times; a third applies himself to the most useless inquiries; that one kills himself by devoting himself to the most frivolous compositions; another in composing the most pedantic works, with no thought for the damage he is doing to himself and how little profit the public will derive therefrom; most do not even have the public in mind, and devour learning in the same way the greedy man devours food, to sate their passion.[57]

What would Tissot have made of Heerkens's description of the hearths that scholars construct in their studies as 'altars of learned Vulcan' (*Docti Mulcibris aris*)—a recherché pun, the author dutifully informs us, on the Roman temple to 'Womanly Fortune' (*Fortunae Muliebris*)?[58] He would probably have diagnosed pathological learning in his Dutch colleague from a perfunctory appraisal of the poem's *mise en page*, where the footnotes regularly crowd out the verse with erudite excursions; into the third book, indeed, he has packed a scholarly dissertation on the causes of Ovid's exile. But many of Heerkens's footnotes are, on closer inspection, as much about his observations of literary and scholarly society in action (including as a participant) as displaying his undeniably considerable book-learning. They are more often a vehicle for purging the poet's personal memories and opinions than, as it were, a receptacle for any obstructive pedantic waste products!

While Heerkens's *De valetudine literatorum* opens on a sombre note—his dismay at the untimely death of so many learned men, and an outpouring of personal grief for the loss of his friend, the over-industrious poet and translator, Justus Conring, at just 23 years of age—our young medical student regards learning, in proper measure, as a good thing. It is true that the 'soldiering of studies' (*militia studiorum*) can be unhealthy: 'I have seen those whom a whole winter spent in their study denied a journey of a hundred steps in the spring; I have seen those who so un-learned sleep through study that it would not return except with the help of

---

[57] '*l'un compile les choses les plus communes, l'autre redit ce qu'on a dit cent fois, un troisième s'occupe des recherches les plus inutiles, celui-ci se tuë en se livrant aux compositions les plus frivoles, celui-là en composant les ouvrages les plus fastidieux, sans qu'aucun d'eux songe au mal qu'il fait, & au peu de fruit que le public en retirera; le plus grand nombre n'a même jamais le public en vuë & ne dévore l'étude que comme le gourmand dévore les viands pour assouvir sa passion*'. Tissot 1769, 139.
[58] Heerkens 1790, 124 and Note.

sleeping drugs'.[59] But Heerkens will not concede that learning is in itself a disease, or even an unhappy life choice:

> There is no-one [in Tissot's book] who would not frighten you away from all study; no-one there you would wish to follow. Believe me, none but a propitious God gives anyone a studious mind, eager for knowledge. He wants, he *wants* to reveal the secrets of his works, and from ancient times he bids you to be wise. Let him [Tissot] laugh at the learned from the Greek and Roman worlds: our exile has returned from this world a cultured man! And there is no field of study that is hateful, none that depresses your spirits, and there is no sad way of life. You can be happy in the company of severe Minerva with a wrinkled brow; you can be happy on Helicon with the happy goddesses [of poetry].[60]

Indeed, learning can even be 'protective': not only do those without culture live less well, they do not live as long. And so Heerkens asks:

> Why is it that, wherever you look, the mob which feeds itself in the fields, or in the city, by the healthy labour of its hands, does not yield a rich crop of old people? Culture is absent! Is it that drink, food, rain, heat and cold, and now work are harmful, now long periods of leisure? And from those who catch the disease that proceeds from these things, an ignorance on a par with disease—an ignorance which was of their life—what, apart from death, can remove it? And so too, though Rome survived long, what wonder is it if, before captured Greece brought culture to that proud race, the earliest citizens rarely reached old age, and almost none to the age of Xenophon?[61]

In an accompanying footnote, Heerkens snarls:

> Tissot may consider this kind of life [sc. of manual labourers] to be healthy, and healthy an ignorance of the ills that proceed from this kind of life. According to Tacitus, those who brought barbarism to Europe lived in this way, the Germans and Goths. [...] Here [referring to his paraphrase from Tacitus *Germania* 15, on the Germans' laziness and delegation of manual work to women and the weak during periods of respite from war], in Tacitus' words, is that famous way of life, which is proof of a great longing for barbarism in those, whether like Tissot, for whom the 'most paradoxical' Rousseau is an object of admiration, or whoever has no regard for the benefits that culture confers.[62]

Tissot is tarred with the same brush, here, as his Swiss compatriot, Rousseau. A discussion of the insalubrious abuse of indoor fires in the third book is the pretext

---

[59] '*Vidi, musaeo quibus omnis bruma peracta,/Ad centum passus vere negabat iter,/Vidi, qui somnum sic dedidicere studendo,/Non nisi somnifera post ut adesset ope*'. Heerkens 1790, 41.

[60] '*Nullus ibi, qui non studio te absterreat omni,/Nullus ibi, cuperes quo praeunte sequi./Crede mihi, mentem studiosam, avidamque sciendi,/Non nisi propitius dat cuicunque Deus./Vult sua, vult operum secreta patere suorum,/Deque aevis veterum te sapuisse jubet./Rideat a Graecis doctos, et ab orbe Latino:/Cultus ab hoc exul redditus orbe fuit./Et nullum studii genus est inamabile, quodque/Tristem animum, vitae triste genusque facit./Laetus apud tetricam caperata fronte Minervam,/Laetus apud laetas sis Helicone Deas.*' Heerkens 1790, 61.

[61] '*Vulgus agris manuum vel in urbe labore salubri/Quod se alit, haud crebros cur dat ubique senes?/Cultus abest. potus, cibus, imber, frigus et aestus,/Jamque nocens labor est, jam diuturna quies?/Morbus et hinc quibus est, par ignorantia morbi,/Quae fuerat vitae, quid nisi fata ferat?/Sic quoque Roma diu cum vixerit, ante superbo/Quam cultum populo Graecia capta dabat,/Quid mirum est, primis si rara Quiritibus aetas,/Et Xenephontaeae par prope nulla fuit?*' Heerkens 1790, 63.

[62] Heerkens 1790, 63, Note.

for a devastating diatribe on the intellectual pretensions of the *parvenu* from Geneva.[63] Heerkens was aware that the most ancient Greek physicians attributed a general decline in human health and longevity to the discovery of fire, and he suggests that this was the likely origin of the myth of Prometheus unleashing disease throughout the earth. In Paris, however, he would discover that the thirteenth-century Catalan physician, Arnaud de Ville-Neuve, had already expressed much the same view, though probably ignorant of the Greek sources. This led Heerkens to approve the ancient teaching all the more, since it was confirmed by Ville-Neuve's experience and not by mere authority, and so he proceeded to promulgate it among the learned men of Paris. But as the doctrine and its double provenance became known, Rousseau began to 'proclaim' (*personare*) it. Heerkens, the younger man, came under suspicion of being a disciple of the older Rousseau, and was warned to keep away from the 'most paradoxical one' (ὰ παραδοζωτατω). Heerkens avers that he was able to live with the fact that:

> [Rousseau] was putting it about that writers not even known to him by name, and indicated to him by *me*, had been rescued from the darkness, read by him, and were most worthy of everyone's attention—but as soon as he abused the information I had provided for the purposes of corroborating his own ill-omened opinions, I fled from his side.[64]

In the continuation of this footnote, Heerkens provides a scathing portrait of Rousseau the *faux savant*, the antithesis of the learned and worthy man of letters who is the poem's de facto addressee:

> Rousseau had come to Paris not long, or at least not very long, before I did; and he had come, like so many of his countrymen, in the hope of making his fortune—but this was a hope given to him by a little knowledge of music, of skilfully painting the notes of songs. He had withal no knowledge of the ancient languages, but a smattering of the scientific terms of French learning common enough among peoples speaking French. That he should become recognised as a philosopher in such a short time he owed to the Procope café, an establishment, that is, near the *Théâtre français*, where the more idle *érudits* used to gather, and among them, too, the most famous *philosophes*. That he who was for a long time their daily auditor should later dare set himself up as their detractor and adversary, he owed to his tongue, which was glib enough, and to the fact that he had equal tickets on his own judgement.[65]

It was not so bad in the beginning, continues Heerkens, when Rousseau merely attacked, albeit with unseemly ignorance and belligerence, the 'more libertine/outspoken philosophers' (*philosophis licentiosioris*). Indeed, he 'alienated none so much at first as those devoted to the field in which he would become famous'. His fame arose from *paradoxes*, which Rousseau's more intelligent opponents pressed him to defend, and that he did most pertinaciously, even convincing *himself* that they were true. As for Heerkens, he says that he was, at first, well enough disposed to the unfortunate Swiss, furnishing him with ancient maxims and counselling him

---

[63] Heerkens 1790, 164–167, N. 36.
[64] Heerkens 1790, 166, Note.
[65] '[...] *linguae debuit satis disertae, et quod de judicio suo haud minus praesumebat*'. Heerkens 1790, 165–166, Note.

not to venture into subjects that were beyond him. But soon Rousseau was not only spouting the most absurd opinions but also publishing them. His verbal and printed assaults 'procured him the hatred of the learned, all of whom deserted this monstrously rude and bad-tempered man. And a comedy soon gave an indication of this odium, the "Badly Educated Man", the *faux savant*, written by him'.[66]

As with his reports elsewhere in the poem on Voltaire and D'Alembert, Heerkens's intellectual appraisal of Rousseau is predicated on his personal experience of his (bad) behaviour. Heerkens takes issue with Rousseau's presumption, his lack of moderation, his stubbornness, and above all, his ungracious treatment of his intellectual peers. The anti-humanism (its 'paradoxicality', in Heerkens, almost synonymous with wanton absurdity) of Rousseau's thought and writing is to be expected from someone who commits such egregious transgressions of the unwritten code of polite conduct within the Republic of Letters.[67] Far from the *learned* man being pathological, the unlearned man becomes, in Heerkens's view, a monster of misanthropy.

As a humanist physician, Heerkens was by no means indifferent to the health effects of the 'passions of the soul'. And so, in the third book of his poem on learned men's health, he gives particular attention to the emotions of scholars. Scholars are prone to anger, which is usually a consequence of their conceit (*fastus*).[68] Heerkens counsels us on how to cultivate indifference to our reputation, how to rise above envy; in short, how to 'man up' and accept criticism. He does this by adducing a series of vivid and affecting examples of scholarly emotions-in-action, gleaned from personal encounters, from the reports of friends and acquaintances, and from his wide reading. Poet and clergyman Jacques Cassagne (1636–1679) was cruelly mocked for his preaching by Boileau, in his third Satire, which led many to suspect that the latter had precipitated Cassagne's madness and premature death. Heerkens judged otherwise: 'd'Olivet may condemn me, France may condemn me: pride, not Boileau, was your undoing, Cassagne!'.[69] A long note supplements these verses, defending, as had Heerkens's own father, Boileau's right to criticise Cassagne's sacred oratory, and furnishing the 'true story' as later repeated to Heerkens in Paris by poet Louis Racine, a friend of the Cassagne family. Cassagne, it seems, had retired from preaching and poetry after Boileau's attacks, but unwisely turned to theology, in which field he published six or seven works. Unfortunately, he was not

---

[66] '[…] *odium fecit eruditorum, qui hominem immaniter incivilem et iracundum destituerunt omnes. odiique huius signum mox dedit comoedia, Male Doctus, le faux savant, ab ipso inscripta*'. Heerkens 1790, 166. Presumably *not* the work of this title by Jacques du Vaure, first performed in 1728. Heerkens may be referring, muddle-headedly, to Rousseau's *Narcisse ou l'amante de lui-même* (1752), the preface of which 'provides some evidence for the view that, whether real or imagined, Rousseau's sense of being persecuted by "adversaries" was anything but an acquisition of old age' (Barber and Forman 1978, 540).

[67] See Bots and Waquet 1997, 113–114 on 'Le savant et la "civilisation des bonnes manières"'.

[68] He perhaps owes this insight to a seventeenth-century Belgian Jesuit, Lieven de Meyere, who had written a three-book Latin poem on Stoic anger management. See Haskell 2011a.

[69] '*Damnet Olivetus, damnet me Gallia: fastus,/Non tua Cassagni Parca, Bolaeus erat*'. Heerkens 1790, 138.

at all suited to theological studies, which he pursued with 'such intemperance' that he soon displayed a 'sick and morose mind to all, though his reason was intact'. Eventually, 'he so alienated his own mind that he was shut up in the hospital of St. Lazare by his family, to finish his life in that sad home'.[70] A melancholy story, worthy of a Tissot—but our cheerful Dutchman promptly pulls a more sanguine one out of the box. Should not Jean Chapelain rightfully have lamented the fate of *his* poem on Joan of Arc, 20 years in the making, which was ultimately published to universal ridicule?[71] Heerkens quips, in one of his more accomplished Ovidian couplets: 'After all that time, Chapelain's long-awaited Girl comes out into the light—an old woman'.[72] Nevertheless, the French poet made a name for himself by ignoring the injury, and was honoured by the whole of France, by Louis XIV no less, and lived some 8 years: 'Pray, poets, for such a spirit when you publish your work!', Heerkens exhorts us, 'this is as honourable as entrusting good writings to fame.'[73]

Those who are susceptible to losing heart, too much afflicted by shame and remorse, are not cut out for a life of learning. On the other hand, the successful man of letters is no impassive Stoic, but will evince a certain 'sensitivity' (*sensus*), and *feel* the emotions he wishes to convey:

> Do you think that those who persuade the general mob, who have been its saving, who were able to advise on the interests of their country, do you think their minds lacked sensitivity, and that they did not grieve themselves whilst others were grieving?[74]

This capacity for feeling is, however, no sign of mental illness: 'The more sensitive the mind is, and the more affected for worthy reasons, the sharper it usually is'.[75] Heerkens gives the example of his esteemed countryman, poet Jacob Cats, who jumped into the grave of his beloved wife.[76] If this behaviour sounds bizarre, Heerkens was far from consigning Cats to a basket of the wretched, mad, and squalid such as Tissot wove; he observes that Cats was both physically well-proportioned and attractive, thus refuting the opinion of Jesuit poet, Jacob Balde, who claimed that a good mind and long life were not consistent with a handsome body.[77]

---

[70] Heerkens 1790, 139, N. 12.

[71] The first 12 cantos of Chapelain's *La Pucelle* were published in 1656.

[72] '*Illa Capellani dudum exspectata Puella,/Post longa in lucem tempora prodit anus*'. Heerkens 1790, 140.

[73] '*Talem animum ostenso vates orate labori!/Tam decet hic, famae quam bona scripta dare*'. Heerkens 1790, 140.

[74] '*Et, qui hominum turbae suadere,salusque fuisse,/Consulere et patriae qui potuere suae,/Horum animos sensu caruisse, nec ex alienis,/Credis et hos propriis non doluisse malis?*'. Heerkens 1790, 144. Heerkens's concept of *sensus* is influenced by his reading of L.-J. Lévesque de Pouilly's *Théorie des sentimens agréables* (Lévesque 1747). He met de Pouilly in Rheims, where he graduated in medicine on his return from Paris. See the preface to Heerkens 1749.

[75] '*Quo sensibilior mens est, affectaque dignis/Quo magis ex causis, acrior esse solet*'. Heerkens 1790, 144.

[76] Heerkens 1790, 145–147.

[77] Heerkens 1790, 160 and N. 30, N. 31.

If we make the comparison with Tissot's oration, we find that the individual historical instantiation of scholarly emotions/passions is much less in evidence. Tissot comments in general terms on the pusillanimity of scholars, on the sadness they suffer on account of their preference for solitude,[78] which leads to misanthropy, dissatisfaction, and 'this disgust with everything, that one can regard as the greatest of ills, since they remove the delight from all the good'.[79] But even where he cites individual men of letters (such as Swammerdam, above), he does so in a rather peremptory fashion, and he certainly does not indulge in anything like the long and digressive biographical anecdotes ubiquitous in Heerkens's poem. Thus Fontenelle is adduced by Tissot as an exemplum of that rare man of letters who reached a happy old age and avoided infirmity, because he mixed 'the pleasures of civilised life with literary work'.[80] Heerkens, too, admires Fontenelle, but his effusions on the senior *philosophe* are, by contrast, much more idiosyncratic and intimate. Take, for example, this charming note on his fondness for Mademoiselle de Scudery:

> Scudery lived ninety-four years, and died on the eleventh of June of the first year of this century—a date I will always remember because, when I approached Fontenelle on that day, at around noon, he told me to follow him to church to pray for Scudery on the anniversary of her death. And while I walked in the street alongside the window of his litter, I was instructed to incline my ear to his face, and he requested of me that, if I survived forty-eight years after his death, I should celebrate the anniversary of his last day as he did of Scudery. And since I had already written these verses, and had arranged for them to be read out to him a few days later, he declared, grabbing my hand, that this token of my affection was pleasing to him, and also because I had considered him worthy to be associated with Fleury and Scudery, through no merit of his own.[81]

We mentioned above Tissot's salutary lesson on the sad fate of Caspar Barlaeus (van Baerle), whose 'excessive studies so weakened his brain, that he believed that his body was made of butter: he avoided fire with care; at the end, weary of his continual terror, he threw himself into a well'.[82] Heerkens's diagnosis of Barlaeus's predicament is more humanistic, and, for that matter, humane. He does not read his fearfulness as a result of intemperate study, but rather as understandable paranoia as a result of real and prolonged religious persecution (for his Arminianism). Heerkens recounts in a footnote the story of a policeman who tailed Barlaeus in the street and snatched from his bag what he took to be an incriminating document, which he duly delivered to the city magistrates. Although the suspected page proved to contain nothing more than a Latin poem on the death of a puppy, Barlaeus never fully recovered from the fright. In any case, Heerkens brings up Barlaeus not as an example of

---

[78] Interestingly, Tissot's friend Johann Georg Zimmerman (1728–1795), physician to King George III, had written an essay advocating the benefits of solitude.

[79] '*ce dégoût de tout, qu'on peut regarder comme les plus grands des maux, puisqu'ils ôtent la jouissance de tous les biens*'. Tissot 1769, 103.

[80] '*les douceurs de la vie civile aux travaux litteraires*'. Tissot 1769, 61–62.

[81] Heerkens 1790, 133, N. 7.

[82] '*études excessives lui affoiblirent tellement le cerveau, qu'il croyoit qui son corps étoit de beure: il fuyoit le feu avec soin; à la fin ennuyé de ses terreurs continuelles, il se précipita dans un puits*'. Tissot 1769, 42.

scholarly excess, but of the perils of becoming embroiled in scholarly disputes. Barlaeus had been unwise enough to respond to silly criticisms made by a certain *haruspex* (sc. 'interpreter') of a short poem of his, and had thereby attracted further controversy and censure. The health of scholars, says Heerkens, will not withstand the psychological stress of such quarrels.[83]

For Tissot, quoting Aretaeus, the 'passion' for learning is all too real—stronger and more dangerous than any other, and overrides love of country, filial and brotherly love, self-love, and even self-preservation.[84] In a long footnote diatribe, Tissot connects a perceived contemporary spike in nervous disease with, among other things, the proliferation of harmful passions (vanity, avarice, ambition, and jealousy), a consequence of the trend towards luxurious city living. But prime among possible causes for this supposed epidemic are the popularisation of the sciences and arts (*'l'amour des Sciences & la culture des Lettres beaucoup plus répandues'*), rampant publication (*'Cette foule de presses qui roulent continuellement en Europe'*), and reading to excess—especially women reading novels.[85] The 'nervous' scholar is thus condemned by association with urban decadence, effeminacy, and moral viciousness.

Heerkens is still living (at least in his head) in a very different world of letters. He assumes a suave Ovidian tone in critiquing the Catonic harshness implicit in the Tissotian/Rousseauian critique of learned culture:

> My Muse does not bid you shun society and culture, nor the human race to be four-footed. So you may know she has not been instructed in the least by the Philosopher of Lake Geneva, she teaches you to stay [in the city], and the origins of the happy life. And all those whom bad education has corrupted in their tender years, and all those whose first youth has been given over to indulgence, she strives, and not in an ill-tempered way, to fortify with advice, and she is satisfied with any attempt at self-discipline.[86]

Tissot's treatise concludes with a rider, anticipating the objection that he has little personal experience or appreciation of the scholarly life, and that he is endorsing the radically negative view of learning advanced by Rousseau in his first *Discourse*. As Vila observes, Tissot proceeds to take

> a moderate stance in the debate then raging over the relative merits and risks of striving to become learned: he argues that the pursuit of knowledge, while not entirely beneficial, can at least be benign to fledgling scholars, as long as they meet certain conditions.[87]

---

[83] And he should know! I discuss the student Heerkens's involvement in a protracted literary quarrel in Groningen in the first chapter of *Prescribing Ovid*. The older Heerkens seems to have felt that he was never forgiven for the satirical indiscretions of his youth (Haskell 2013).

[84] Tissot 1769, 252–253.

[85] Tissot 1769, 199–202. The footnote, which runs over four pages, begins: '*Les maladies des nerfs sont beaucoup plus fréquentes & plus varies qu'elles ne l'étoient il y a soixante ans*'.

[86] '*Non mea convictum Musa, aut contemnere cultum,/Humanum quadrupes nec jubet esse genus:/Deque Sopho nihil ut videatur docta Lemani,/Restare, et felix vita sit unde, docet:/Et quos corrupit tenerum mala cura per aevum,/Et data deliitiis prima juventa suis,/Consilio cunctos studet haud morosa tueri,/Contentam quovis seque Catone, probat*'. Heerkens 1790, 166–167.

[87] Vila 1998, 103.

These conditions make of the scholar a

> type of patient who must submit to constant control, not only physical control [...], but also moral control, which Tissot exerts by exhorting his readers to cultivate the arts and sciences in a manner that is cool-headed, self-disciplined, and socially acceptable.[88]

Given the vehemence and peculiar inflections of his tirade against contemporary nervous disease, one wonders whether Tissot was ever as concerned with the older, predominantly masculine and Latinate, Republic of Letters—to which, in many ways, he still belonged—as with a new, more democratic, literary culture of novels, translations, and popularisations 'for the ladies', with its feared moral and social consequences.[89] Be that as it may, Tissot's view on what we would now call 'life-long learning' is that it is, for most people, superfluous, if not downright dangerous. Thus he advises not only against embarking on studies too young, but also in middle age, and he warns against increasing the pace of our studies or venturing into unfamiliar fields.[90]

Heerkens, on the other hand, might appear to be verging in the direction of the views advanced by Tissot's Parisian foils, Le Camus and Vandermonde, who advocated the medical enhancement of sensibility and the cultivation of learning for self- and societal improvement.[91] I suspect, though, that Heerkens was less interested in any project of Enlightenment eugenics than in renewing hope in those, like himself, who wished to persevere in an older style of intellectual life that was fast becoming obsolete. In contrast to the Rousseauian/Tissotian caricature of obsessional and solitary erudition, Heerkens's ideal life of learning is an eminently social activity: 'How great is Ménage among the learned! And he confesses that he learned more from company than from his books'.[92] This life certainly has its psychological and physical dangers, but also its very real emotional compensations. Heerkens charges us to learn from *everyone* that God puts in our path, including the 'bad, stupid, proud, and harsh'; but

> get to know more, know very well your peers in your field, whether your homeland or a journey has given you the opportunity to visit them. And select from those who are fashioned

---

[88] Vila 1998, 103.

[89] Charlotte Lennox's *The Female Quixote* (1752) has a young protagonist whose pathological reading makes her believe she is living in a novel, and so requires medical treatment. That fear is also expressed in many contemporary reviews of Gothic romances. I owe these observations to Karin Kukkonen, St. John's College, Oxford.

[90] It is ironic that Tissot advises mature learners against just that sort of novel mental activity which is advocated by some modern gerontologists and psychologists for preserving memory function and improving quality of life (including learning a new language!): '*Les nouvelles idées dont il s'occupent, mettent nécessairement en action de nouvelles fibres dans le cerveau pour lequel cela forme un état violent qui affoiblit le genre nerveux. J'ai connu un très habile Théologien qui ruina absolument sa santé en suspendan ses études habituelles pour se livrer à celle de l'hébreu*' (Tissot 1769, 129).

[91] See Vila 1998.

[92] '*Quantus apud doctos Menagius! Isque fatetur,/Se plus convictu, quam didicisse libris*'. Heerkens 1790, 41–42, and N. 60.

after your own temperament, and from the peaceful ones, those whom you might wish to follow in all their actions.[93]

In this context, Heerkens favours Ovid, and, in his century, Fontenelle, whom he 'keeps before his eyes if ever quarrels come'.[94]

While the best minds are, Heerkens concedes, 'driven by more nerves',[95] the informed scholar will take his exercise and recreation, and will enjoy the society—including the virtual society—of learned friends:

> For what Moor or Indian, or whosoever dwells in the wild world of America, ever lived alone with his wandering wife? The lands of Mexico and Peru, and the Moor, yielded happy companions before the arrival of the Spanish ships. But my race [sc. *gens de lettres*] has given me to know companions far more blessed, and from every quarter, than those whom I see and venerate [sc. in person]. There is no room for enmity where great culture, in every word, teaches friendships, and hearts to be pious. Let him come here, he who labours with a mind injured by studies—from these men he will learn that no-one is harmed by his study![96]

Acknowledgments I acknowledge the support of the Australian Research Council. Part of the present chapter has now appeared in Haskell 2013, Chap. 2.

References

Aretaeus. 1856/1972. *The Extant Works of Aretaeus, The Cappadocian*. Trans. Francis Adams. Boston: Milford House.
Barber, B.R., and J. Forman. 1978. Introduction to Jean-Jacques Rousseau's 'Preface to Narcisse'. *Political Theory* 6(4): 537–542.
Bots, Hans, and Françoise Waquet. 1997. *La République des Lettres*. Paris/Bruxelles: Belin/De Boeck.
Lévesque de Pouilly, L.-J. 1747. *Theorie des sentimens agreables*. Geneva.
Ferrand, Jacques. 1991. *Treatise on Lovesickness*. Trans. and ed. Donald A. Beecher and Massimo Ciavolella. Syracuse: Syracuse University Press.
Ficino, Marsilio. 1489. *De vita libri tres*. Florence.
Fischer, K.-D. 1988a. Das Gesundheitsgedicht des Burkhard von Horneck (†1522). *Swiss Journal of the History of Medicine and Sciences* 45(1): 31–48.
Fischer, K.-D. 1988b. Medici poetae de sanitate conservanda. *Vox Latina* 42(94): 472–485.
Ford, Philip. 1999. Claude Quillet's *Callipaedia* (1655): Eugenics treatise or pregnancy manual? In *Poets and teachers: Latin didactic poetry and the didactic authority of the Latin poet from the Renaissance to the present*, ed. Yasmin Haskell and Philip Hardie, 125–140. Bari: Levante.

---

[93] '*Nosce magis, studiique tui pernosce coaevos,/Patria, visendos seu via facta dabit./Deque tuos factis ad mores, deque quietis/Selige, quos cupias cuncta per acta sequi*'. Heerkens 1790, 213.

[94] '*Et lis, ante oculos, si mihi fiat, erunt*'. Heerkens 1790, 213.

[95] '*nervisque a pluribus acta*' Heerkens 1790, 213.

[96] '*Nam quis solivaga cum conjuge Maurus et Indus/Degit, et Americi quisquis in orbe fero est?/Felices socios Peruaeque et Mexima tellus./Maurus et Hispanas protulit ante rates./Sed longe socios magis omni a parte beatos./Quam video, et veneror, gens mihi nosse dedit./Non locus offensae est, ubi verbo multus in omni/Cultus amicitias et pia corda docet./Huc veniat, laesa studiis qui mente laborat./His studium nulli cernet obesse suum*'. Heerkens 1790, 214–215.

Gowland, Angus. 2006a. The problem of Renaissance Melancholy. *Past and Present* 191(1): 77–120.
Gowland, Angus. 2006b. *The worlds of Renaissance Melancholy: Robert Burton in context.* Cambridge: Cambridge University Press.
Haskell, Yasmin. 2007a. A Dutch doctor's observations on the health of scholars, young and old: Gerard Nicolaas Heerkens' *De valetudine literatorum* (Leiden and Rheims, 1749; Groningen, 1790). In *Miraculum eruditionis: Neo-Latin studies in Honour of Hans Helander*, eds. Maria Berggren and Christer Henriksen, 151–166. Uppsala: Almqvist & Wiksell.
Haskell, Yasmin. 2007b. Poetry or pathology: Jesuit hypochondria in early modern Naples. *Early Science and Medicine* 12(2): 187–213.
Haskell, Yasmin. 2008. Latin poet-doctors of the eighteenth century: The German Lucretius (Johann Ernst Hebenstreit) versus the Dutch Ovid (Gerard Nicolaas Heerkens). *Intellectual History Review* 18(1): 91–101.
Haskell, Yasmin. 2011a. Early modern anger management: Seneca, Ovid, and Lieven de Meyere's *De ira libri tres* (Antwerp, 1694). *International Journal of the Classical Tradition* 18(1): 36–65.
Haskell, Yasmin. 2011b. The anatomy of hypochondria: Malachias Geiger's *Microcosmus hypochondriacus* (Munich, 1652). In *Diseases of the imagination and imaginary disease in the early modern period*, ed. Yasmin Haskell, 271–295. Turnhout: Brepols.
Haskell, Yasmin. 2013. *Prescribing Ovid: The Latin works and networks of the enlightened Dr Heerkens.* London/New York: Bloomsbury Academic.
Heerkens, Gerard Nicolaas. 1749. *De valetudine literatorum.* Leiden and Rheims.
Heerkens, Gerard Nicolaas. 1751. *Satire to Cornelius Valerius Vonck.* Groningen.
Heerkens, Gerard Nicolaas. 1790. *De valetudine literatorum.* Groningen.
Hummel, Pascale. 2002. *Moeurs erudites: Étude sur la micrologie littéraire (Allemagne, XVIe-XVIIIe siècles).* Geneva: Droz.
Kummel, W.F. 1984. Der *Homo litteratus* und die Kunst, gesund zu leben: Zur Entfaltung eines Zweiges der Diätetik im Humanismus. In *Humanismus und Medizin*, ed. R. Schmitz and G. Keil, 67–86. Weinheim: Acta Humaniora.
Mikkeli, Heikki. 1999. *Hygiene in the early medical tradition.* Helsinki: Academia Scientiarum Fennica.
Schmidt, Jeremy. 2007. *Melancholy and the care of the soul: Religion, moral philosophy, and madness in early modern England.* Aldershot: Ashgate.
Sorkin, David. 2008. *The religious enlightenment: Protestants, Jews, and Catholics from London to Vienna.* Princeton: Princeton University Press.
Tissot, Samuel Auguste André David. 1769. *De la santé des gens de lettres.* Lausanne.
Vila, Anne C. 1998. *Enlightenment and pathology. Sensibility in the literature and medicine of eighteenth-century France.* Baltimore: Johns Hopkins University Press.

Chapter 7
Penseurs Profonds: Sensibility and the Knowledge-Seeker in Eighteenth-Century France

Anne C. Vila

Abstract The best-known intellectual persona of the French Enlightenment, the *philosophe*, is typically associated not with the vicissitudes of sensory, corporeal existence, but with reason, truth-telling, and the pursuit of social and political reform. However, like many other aspects of eighteenth-century culture, the figure of the thinker was deeply inflected by sensibility's rise as a concept that bridged body, mind, and milieu. This chapter focuses on the absorbed thinker as a type to reconstruct what sensibility was held to do in the mind and body during the act of intense cerebration. It examines the ambiguous affective and sensory state which various moralists and physicians ascribed to thinkers observed or imagined in the state of absorption. It then considers some of the purposes to which Denis Diderot put the figure, focusing particularly on the absentminded geometer characters that appeared in his fictional dialogue *Le Rêve de d'Alembert* (1769) and in the *Eléments de physiologie* (1778). Finally, it considers what those depictions imply, both for Diderot's views on the thinking process and for existing historiographical accounts of sensibility in the Enlightenment era.

> There are no deep thinkers, no ardent imaginations that are not subject to momentary catalepsies. A singular idea comes to mind, a strange connection distracts us, and our heads are lost. We come back from that state as from a dream, asking those around us, 'where was I? What was I saying?'[1]

Denis Diderot's bemused fascination for 'deep thinkers' and 'ardent imaginations' reflected both his own, occasionally idiosyncratic, views on human nature

---

[1] '*Point de penseurs profonds, point d'imaginations ardentes qui ne soient sujets à des catalepsies momentanées. Une idée singulière se présente, un rapport bizarre distrait, et voila la tête perdue on revient de là comme d'un rêve*: *on demande à ses auditeurs, où en étais-je? Que disais-je?*'. Diderot 1778/1975–, 328–329. All translations are my own, unless otherwise noted.

A.C. Vila (✉)
Department of French and Italian, University of Wisconsin-Madison,
618 Van Hise Hall, 53706 Madison, WI, USA
e-mail: acvila@wisc.edu

and some of the larger currents of his era. Thinkers were widely celebrated in the eighteenth century: geniuses were venerated, and intellectuals in general enjoyed greater social prominence.[2] However, despite widespread efforts to bring the life of learning into closer alignment with the practices and values of polite society, an aura of difference—strangeness, even—surrounded the knowledge-seeker as a type. This was not simply because some intellectuals remained wilfully aloof from *le beau monde*, as Jean d'Alembert recommended in his *Essai sur la société des gens de lettres et des grands* (1753).[3] It was also due to the pervasive belief that the true 'deep thinkers' of the world were constituted differently from the non-intellectual cultural elite (as well as from the common herd). According to this view, those who devoted themselves fully and intently to learned endeavour had unique ways of feeling and sensing—including, in Diderot's estimation, an odd tendency to slip in and out of 'catalepsies' when they were gripped by an idea.

Approaching thinkers from the angle of sensibility may seem odd in itself, given that the best-known intellectual persona of the day, the Enlightenment *philosophe*, is typically associated not with the vicissitudes of sensory, corporeal existence, but with reason, truth-telling, and the pursuit of social and political reform.[4] However, like many other aspects of eighteenth-century culture, the figure of the thinker was deeply inflected by sensibility's rise as a concept that bridged body, mind, and milieu.

Various factors were involved in both the emergence of sensibility and the embodied view of knowledge-seeking it inspired. These included the revalorisation of sentiment and the passions in European moral philosophy and literature, the emphasis which philosophers like Etienne Bonnot de Condillac placed on sensations in the formation of knowledge and subjectivity, and the shift towards a more physiological conception of the common sensorium or 'seat' of the soul.[5] The biomedical sciences also played a key role: in the 1740s, the Swiss physician Albrecht von Haller published ground-breaking experimental investigations on the reactive properties of muscles and nerves, which highlighted the inadequacies of mechanistic explanations of the body's physiological processes, and proposed the more dynamic notions of irritability and sensibility to replace iatromechanistic models.[6] Within French medicine, the most important response to Haller came from the vitalist physicians and graduates of the Montpellier medical faculty, starting with *Recherches anatomiques sur la position des glandes* (1752), in which Théophile de

---

[2] See Bonnet 1998; Bell 2001, 107–139.

[3] Lorraine Daston argues that the Enlightenment intellectual embraced 'an ideology of distance, both metaphorical and literal, from all human ties' (Daston 2001, 121).

[4] See Condren et al. 2006; Wilson 2008; Brewer 2008, 49–74.

[5] On the revalorisation of the passions within moral philosophy, see Cook 2002. On the rise of sensibility in French literature, see (among many sources) Vila 1998, from which some of the following discussion is adapted. Karl Figlio offers an incisive account of the ways in which several key theorists integrated psychological/philosophical notions of the mind into investigations of the physiological/anatomical aspects of the nervous system in Figlio 1975.

[6] See Steinke 2005.

Bordeu offered a vision of the living body as a federation of semiautonomous sensitive parts, held together both by the nervous system and by the influence of the three major vital centres or 'departments' (the heart, the stomach, and the brain). New theories of psychology, as well as of physiology, arose in the wake of Haller's work, including some that led in the direction of monism—as, for example, in Julien Offray de la Mettrie's *L'Homme machine* (1747).

The model of thinking that emerged in the French Enlightenment thus wove together strands from various sources. At its heart was a theory that emphasised the fundamental similarity of all modes of sensory receptiveness, internal as well as external—and that supposed complex entanglements among the various parts of the human being. Thinking was a holistic process involving not just the brain, but also other physiological centres like the abdomen, and it had profound, sometimes strange, effects on the senses and consciousness.[7] This view of intellectual activity was distinct both from the paradigm of the immaterial Cartesian cogito which preceded it and the paradigm of 'brainhood' that developed later.[8]

As Alexander Cook has noted, the eighteenth century 'witnessed an unprecedented boom in literature devoted to exploring or theorising the mechanisms of human sensibility'.[9] Given that this literature covered a wide range of genres, Cook proposes that we approach it by adopting one of the strategies of differentiation that were common among theorists of the time (for example, the distinctions that they themselves drew amongst different sorts of feeling). In that spirit, I will borrow the practice of typology that was used in several genres to identify distinct types of sensibility across the human spectrum. Typological thinking about sensibility underpinned the creation of various cultural personae of the French Enlightenment: the vaporous woman, the man of refined aesthetic judgment, the dispassionate actor, the apathetic Sadian master libertine, and the knowledge-seeker. The mechanisms of sensibility followed peculiar paths in the last of those personae, for reasons that had as much to do with the distinct temperament ascribed to cerebralists as with the period's styles of intellectual self-fashioning.

The aim of this chapter is to use the figure of the absorbed thinker as a means of reconstructing what sensibility was held to do in the mind and body during the act of intense cerebration. It will examine the ambiguous affective and sensory state which various writers ascribed to thinkers observed or imagined in the state of absorption. It will then consider some of the purposes to which Diderot put the figure, focusing particularly on the absentminded geometer characters that appeared in his fictional dialogue *Le Rêve de d'Alembert* (1769) and in the *Eléments de physiologie* (1778). Finally, it will consider what all of this implies, both for Diderot's approach to the property and for existing historiographical accounts of sensibility in the Enlightenment era.

---

[7] On sensibility's association with susceptibility to external stimuli, see Janković 2010, 15–40. On Montpellier medical vitalism, see Williams 1994 and 2003; Rey 2000; Kaitaro 2007; Wolfe and Terada 2008.

[8] See Alberti 2009; Vidal 2009.

[9] Cook 2009, 457.

7.1 The Pleasures and Dangers of Intellectual Absorption

In his *Encyclopédie* article 'Etude', Louis de Jaucourt drew on a long humanist tradition that regarded the pleasures of study as the highest, most universally rewarding source of human contentment.[10] Citing Cicero for support, he declared that the contemplative life was fully compatible with the values and duties of active life—adding that, rather than clinging to old stereotypes and treating scholars with mocking disdain, the social elite of his day should recognise the benefits that study could have for them personally, as well as for the nation and humanity at large. The moral effects which Jaucourt attributed to study—admiration for true glory, zealous love of country, and enhanced sentiments of humanity, generosity, and justice—illustrate the centrality of feeling in this period's view of intellectual endeavour.

Like study, sensibility was seen as an enhancing quality. This is evident in Jaucourt's short *Encyclopédie* entry 'Sensibilité (Morale)', where he defined sensibility as 'a tender and delicate disposition of the soul that makes it easily moved or touched [...] Sensitive souls have more existence than others: good things and bad are multiplied in them.'[11] Sensibility was thus a trait that magnified feeling and made the sensitive more humane, more empathetic, and more intelligent; on the other hand, it might also multiply their negative qualities or experiences. That double-edged perspective was reflected elsewhere in the *Encyclopédie*: in 'Digestion' and 'Vapeurs', intensified feeling was attributed to people who constantly and fretfully observed their physical sensations, a group that included *gens de lettres* along with aristocrats, ecclesiastics, *dévots*, women of leisure, and people worn out from debauchery.[12] And in the medical entry 'Sensibilité, Sentiment', the Montpellier-trained physician Henri Fouquet equated heightened sensitivity in one body part with disruption of the overall animal economy.

Fouquet's article is revealing on both a conceptual and a semantic level. First, he characterised sensibility as a 'physical or material passion' common to all animals, which allowed individual organs to perceive and respond to the impressions made by external objects.[13] Second, evoking the theory of vital centres which Louis de

---

[10] '*L'étude est par elle-même de toutes les occupations celle qui procure à ceux qui s'y attachent, les plaisirs les plus attrayans, les plus doux & les plus honnêtes de la vie; plaisirs uniques, propres en tout tems, à tout âge & en tous lieux. Les lettres, dit l'homme du monde qui en a le mieux connu la valeur, n'embarrassent jamais dans la vie; elles forment la jeunesse, servent dans l'âge mûr, & réjoüissent dans la vieillesse; elles consolent dans l'adversité, & elles rehaussent le lustre de la fortune dans la prospérité; elles nous entretiennent la nuit & le jour; elles nous amusent à la ville, nous occupent à la campagne, & nous délassent dans les voyages*: Studia adolescentiam alunt. [...] Cicer. pro Archia'. (Jaucourt 1756, 86.)

[11] '*une disposition tendre & délicate de l'ame, qui la rend facile à être émue, à être touchée [...] Les ames sensibles ont plus d'existence que les autres: les biens & les maux se multiplient à leur égard*'. (Jaucourt 1765, 52.)

[12] The Montpellier-trained physician Gabriel Venel implied that those who fretted over petty ailments like *digestion fougueuse* suffered mainly from self-absorption: he called them '*les gens qui s'observent ou qui s'écoutent*'. (Emphasis in original. Venel 1754, 1002.)

[13] Fouquet 1765, 40.

Lacaze (uncle of Bordeu) had sought to popularise in his *Idée de l'homme physique et moral* [1755]), Fouquet posited that the epigastric region acted as a sort of fulcrum or rallying centre for many, if not all, of these organic passions.[14] Finally, he described organic sensibility as a 'taste' or tact that could turn it in either of two opposing directions: an expansive 'intumescence' that was triggered by positive, pleasing stimuli; or a compression incited by negative ones.[15] A compression was a crisis, in the medical sense of a process that moved from irritation, to climactic reaction, to resolution: an organ reacting to an unpleasant stimulus would recoil until its sensitive principle came back to 'consciousness' and expelled the humours that it had concentrated within itself—affecting, for good or bad, all the organs in its vicinity. Sensibility's overall physiological scheme thus entailed an intricate interplay between the particular organs or vital centres within the body, each of which felt its own passions and expanded or compressed in reaction to them. Vital departments were more or less lively depending on how much stimulation they got—that is, on how much sensibility was 'transported' to them—as a result of habit, age, sex, climate, and other factors.[16]

As Fouquet's text illustrates, the medical vocabulary used to explain sensibility was suffused with psychological metaphors, a rhetorical technique that lent an air of dynamic agency to the workings of the organs inside the body. Human beings were, in this view, teeming with passions, pleasures, and pains deep within themselves, whether they realised it or not; and the more they stimulated certain vital centres— the brain, the heart, the stomach, and so on—the more those parts developed their own tastes, needs, and sensitivities. Out of this theory, medical theorists spun a functional anthropology that categorised people according to the organ or vital centre that dominated their existence.

The tendency to set *gens de lettres* apart as a group was clearly tied to this biomedical effort to typologise human beings along differential lines.[17] It was also connected to the period's veneration for great thinkers, which produced an abundance of eulogistic and biographical literature on France's most eminent philosophers, scientists, and literary writers—much of it built upon the notion that true geniuses possessed a special, brain-centred constitution. In some cases, brain-centredness was equated with tepidness in the affective realm: as Madame de Tencin put it while pointing at the chest of Bernard Le Bovier de Fontenelle, 'what you've got there is all brain', thereby echoing the widespread impression that Fontenelle was a cold fish, indifferent to the tender-hearted sensibility then in vogue.[18] More typically, however, this constitution was endowed with its own kind of emotional intensity. Fontenelle himself recounted that Malebranche was seized at the age of 26 with a life-changing passion for reading Descartes when he

---

[14] Fouquet 1765, 42.
[15] Fouquet 1765, 41–42.
[16] Fouquet 1765, 51.
[17] See Williams 1994, 50–62.
[18] '*C'est de la cervelle que vous avez là*'. Cited by Pierre Moreau in Moreau 1960, Vol. 1, 465.

stumbled upon the *Traité de l'Homme* in a Parisian bookstore.[19] Passion of this sort was central to the foundational story which biographers often told of a great thinker's discovery of his/her intellectual vocation, as was the theme of disdain for health and neglect of the body.

Ardour for learning went hand in hand with a penchant for seclusion. Partial retreat from the world had long been central to the group *habitus* of European intellectuals, which developed when fifteenth-century Northern European scholars moved from university or monastic settings into urban family households, creating cloister-like spaces within them that functioned, as Gadi Algazi has put it, as a 'shield for a scholar's vulnerable self'.[20] What the eighteenth century added was an updated list of the dangers to which the scholarly self was held vulnerable: the greatest dangers came not from the world outside the scholar's study, but from the engrossing activities conducted within.

Moralists who emphasised the social mission of learning cautioned intellectuals that they might become misanthropic and detached if they spent too much time confined with their books.[21] Others worried more about the extreme absorption induced by intense mental application. In Condillac's view, the fault lay with the imagination, which sometimes prompted the mind to shut itself off from even the most pressing information coming from the external world via the senses.[22] Citing the famous case of Archimedes, the ancient mathematician who was too lost in thought during the Roman siege of Syracuse in 212 BC to notice that his life was in danger, Condillac depicted deep thinkers as the group most liable to lose touch with the real world and to heed only the kind of attention caused by the imagination, 'whose characteristic is to arrest the impressions of the senses in order to substitute for them a feeling independent of the action of external objects'.[23]

Archimedes was, in fact, frequently cited in Enlightenment-era discussions of intellectual absorption, perhaps due to the enduring popularity of Plutarch's *Lives* among educated readers. Plutarch's life of Marcellus included two accounts of Archimedes in contemplative oblivion. The first was the tale that

> the charm of his familiar and domestic Siren made him forget his food and neglect his person, to that degree that when he was occasionally carried by absolute violence to bathe or have his body anointed, he used to trace geometrical figures in the ashes of the fire, and

---

[19] On Fontenelle's account of Malebranche's passionate reading of Descartes, see Ribard 2003, 117–19.

[20] Algazi 2003, 26.

[21] Louis-Sébastian Mercier, for example, waxed lyrical about the delights enjoyed exclusively by cerebralists, but also warned that the attraction of reading was liable to turn some into solitary misanthropes. See Mercier 1764, 1766.

[22] '*Le pouvoir de l'imagination est sans bornes. Elle diminue ou même dissipe nos peines, et peut seule donner aux plaisirs l'assaisonnement qui en fait tout le prix. Mais quelquefois c'est l'ennemi le plus cruel que nous ayons: elle augmente nos maux, nous en donne que nous n'avions pas, et finit par nous porter le poignard dans le sein*'. Condillac 1746/1973, 147.

[23] '*dont le caractère est d'arrêter les impressions des sens, pour y substituer un sentiment indépendant de l'action des objets extérieurs*'. Condillac 1754/1984, 30.

diagrams in the oil on his body, being in a state of entire preoccupation, and, in the truest sense, divine possession with his love and delight in science.[24]

The second was the story of his demise, when he was so 'intent upon working out some problem by a diagram, and having fixed his mind alike and his eyes upon the subject of his speculation' that he either failed to notice or ignored the Roman soldier who had been sent to take him to appear before General Marcellus—so enraging the soldier that he killed Archimedes instantly.[25]

The second anecdote regarding Archimedes inspired a variety of applications in eighteenth-century texts on the pleasures and dangers of mental absorption. Julien Offray de La Mettrie used it in his dedication to *L'Homme Machine* (1747) to paint an erotically tinged picture of the 'ecstasies' of knowledge-seeking. The Encyclopedist Fouquet mentioned the Archimedes story while observing that the suspension of the senses triggered by deep meditation was similar to that created by pathological conditions like melancholy and mania.[26] In his *Encyclopédie* article 'Attention', Yvon gave the tale a more benign spin, encouraging readers to emulate famous historical people who possessed great powers of intellectual concentration, even when the world around them was being sacked.[27]

Clearly, theories varied on what was happening to the thinker in these moments. For some, Archimedian attention exemplified optimal mental concentration, the state achieved by those rare souls capable of enjoying the sublime bliss of a meditative trance. This was the view of Yvon, who favoured blocking out sensations as much as possible to focus the mind on the quest for truth.[28] It was also the view of naturalist Charles Bonnet who, as Lorraine Daston notes, erected a veritable cult around painstaking focus on single objects of study.[29] Such defences coexisted, however, with concern over the mind-consciousness split that seemed to occur during full absorption—a split whose operations were mysterious and sometimes troubling, given the apparent absence of voluntary regulation and direction.[30]

7.2 Medical Views on *'Penseurs Profonds'*

Enlightenment physicians also spent a good deal of time contemplating the peculiar temperament and behaviour of cerebralists. From the 1750s onward, intellectuals were a distinct patient group held to suffer from nervous constitution, poor hygiene, and unhealthy work habits. As Charles Augustin Vandermonde declared in his entry

---

[24] Plutarch 1683–1686/2008, 484.
[25] Plutarch 1683–1686/2008, 485.
[26] Fouquet 1765, 46.
[27] Yvon and Formey 1751, 842–843.
[28] Yvon and Formey 1751, 840–841.
[29] Daston 2004.
[30] On related concerns in the British context, see Sutton 2010.

on 'Maladies des gens de lettres' in the *Dictionnaire portatif de santé* (1759), when intellectuals chased after the 'flattering' pleasure of discovering truths, they strained their nerves beyond their natural capacity and harmed the nervous spirits. He cited as proof the heaviness and weakness which scholars commonly felt when they had worked too much, as well as their reddened, inflamed faces.[31]

Medical warnings about the reckless pursuit of intellection abounded during this period. As the Swiss physician Johann Georg Zimmermann put it in his influential *Treatise on Experience in Physic* (original German edition 1763), 'the desire to acquire enlightenment or to make use of the knowledge which one has acquired can easily be ranked among the passions, because it is so strong in some people that it absorbs almost all of their other passions'.[32] Those who applied their minds too intently were, he emphasised, susceptible to numerous ailments, including digestive disorders, debilitating headaches, weakened nerves, hypochondria, loss of sight and hearing, and profound melancholy.

The most developed argument on the dangers of overstudy was put forth by Zimmermann's compatriot and friend Samuel-Auguste Tissot in *De la Santé des gens de lettres* (first edition 1768; expanded third edition 1775). On the one hand, Tissot took a dim view of overzealous scholars, declaring that they were 'like lovers who fly off the handle when one dares to say that the object of their passion has defects; moreover, they almost all have the sort of fixity in their ideas that is created by study'.[33] Yet on the other hand, he offered a host of therapies for study's debilitating health effects—even in cases that involved strange sensory and nervous impairments.

Throughout this book, Tissot emphasised that, when pursued to excess, mental application did serious harm to virtually every body part. These included the sense organs and nerves, whose maladies he would soon catalogue more systematically in his *Traité des nerfs* (1778–1780). One of the cases he discussed involved an English gentleman who had consulted Tissot to report that he had gotten so engrossed in mathematics that he had lost the use of his eyes and eventually his brain, despite showing no signs of physical impairment.[34] Tissot also cited the case (borrowed from Zimmermann) of a 'young Swiss gentleman [...] who buried himself in the study of Metaphysics, and soon felt a mental weariness which he combatted with new efforts of application'. After 6 months of even more intense intellectual efforts, the young man's ailment became so severe that 'his mind and

---

[31] Vandermonde 1759/1760, Vol. 2, 80–81.

[32] '*L'envie d'acquérir des lumières, ou de faire usage des connaissances que l'on a acquises peut sans difficulté se ranger parmi les passions, puisqu'elle est si forte dans quelques personnes, qu'elle y absorbe presque toutes les autres passions*'. Zimmermann 1774/1855, Vol. 3, 477.

[33] '*Ils sont comme les amants qui s'emportent quand on ose leur dire que l'objet de leur passion a des défauts; d'ailleurs ils ont presque tous cette espèce de fixité dans leurs idées que donne l'étude*'. Tissot 1775, 132.

[34] '*J'ai été consulté par un gentilhomme anglais qui, étant à Rome, se livra si fort à l'étude des Mathématiques qu'au bout de quelques mois il ne pût plus se servir de ses yeux quoiqu'on n'y remarquât aucun vice extérieur*'. Tissot 1775, 21–22.

senses fell gradually into the most complete state of stupor'.[35] The doctors treating him feared he was incurable, but managed to devise a method for restoring the normal functioning of his sense organs: this consisted in having someone stand very close to the patient and read a letter in a thundering voice, which woke him up painfully, thereby unblocking his ears.[36] The therapy was continued over the course of a year until all of his senses were restored, and the fellow went on to become 'one of our best Philosophers'.[37]

Such case histories read almost like an inversion of the popular sensationalist fable (put forth by Condillac, among others) of the statue who came to life through the successive activation of the sense organs: whereas the statue enjoyed expansive sentience starting with the smell of the rose, overzealous scholars went cataleptic. Although some doctors used the term 'ecstasy' to characterise that state, the portraits they offered of people in the grips of deep cerebration were far from exalting. Take, for example, this case from the third, expanded edition of *De la santé des gens de lettres*:

> If one considers a man plunged in meditation, one sees that all the muscles of his face are stretched; they even seem at times to be in convulsion; and in the lovely preface which he added to the English translation of this work, Mr. Kirkpatrick cites a fact that must find a place here: 'I knew,' says he, 'a gentleman with a very active genius who, when he thought intensely, had all the fibres of his forehead and a part of his face as visibly agitated as the chords of a harpsichord that is being played in a very lively manner.'[38]

In rhetorical terms, this sort of portrait had a simple purpose: to alarm. To that end, physicians like Tissot used techniques reminiscent of novels like Montesquieu's *Les Lettres persanes* and Graffigny's *Les Lettres d'une Péruvienne*, where familiar things like city streets or scissors were transformed into bewildering objects when

---

[35] '*Mon ami M. ZIMMERMAN, rapporte un autre exemple de l'épuisement litteraire trop intéressant pour l'omettre ici : Un jeune gentilhomme Suisse, dit cet habile Médecin, donna tête baissée dans l'étude de la Métaphysique, bientôt il sentit une lassitude d'esprit, à laquelle il opposa de nouveaux efforts d'application, ils augmenterent la foiblesse, & il les redoubla. Ce combat dura six mois, & le mal augmenta au point que le corps & les sens s'en ressentirent. Quelques remèdes rétablirent un peu le corps, mais l'esprit & les sens tomberent par une gradation insensible dans l'état de stupeur le plus complet. Sans être aveugle il paroissoit ne pas voir; sans être sourd il paroissoit ne pas entendre; sans être muet il ne parloit plus*'. Tissot 1775, 22–23.

[36] Tissot 1775, 23.

[37] Tissot 1775, 23.

[38] '*Si l'on considère un homme plongé dans la méditation, on voit que tous les muscles de son visages sont tendus; ils paroissent même quelquefois en convulsion; et M. Kirkpatrick cite, dans la belle préface qu'il a mise à la tête de la traduction angloise de cet ouvrage, un fait qui doit trouver place ici. 'J'ai connu, dit-il, un gentilhomme d'un génie fort actif, qui, quand il pensait fortement, avoit toutes les fibres de son front, et d'une partie de son visage, aussi visiblement agitées, que les cordes d'un clavecin dont on joue très vivement.*' Tissot 1775, 14–15. Curiously, Tissot altered the sex of the person depicted by Kirkpatrick from female to male: Kirkpatrick originally remarked in his 'Annotator's Preface' that 'I have known a gentlewoman of a most active mind, who, when intensely thinking, had all the nervous filaments of her forehead, and part of her visage, as visibly twitched and agitated, as the wires of a harpsichord are, when vibrating some sprightly air in music'. (Tissot 1769, xx–xxi.)

seen through a foreigner's eyes. They also employed analogies that echoed the contemporary fascination with machines and automata.[39] Their aim was to offer their scholarly readers both an unsettling mirror in which to see themselves and some vivid lessons on the physiological or pathological mechanisms which deep thinking seemed to set in motion.

7.3 Sensibility, Machinality, and 'Deep Thinking' in Diderot

Denis Diderot was also fascinated by the strange processes that intense thinking seemed to involve. That fascination was driven in part by his materialist ideology: intent on externalising the secrets of nature and demolishing the notion of an immaterial soul, he undertook to objectify everything from generation to the faculties of the mind and envision them all as processes that arise out of natural organisation.[40] Another factor was Diderot's rejection of iatromechanistic theories: inspired by the Montpellier vitalist doctors, he stressed the irreducible nature of all vital phenomena and the holistic interaction of higher and lower levels of organisation.[41] However, Diderot did not reject all mechanical explanatory models; to the contrary, he often found it useful to approach the processes of sensing, feeling, and thinking as mechanisms, and he was just as likely to deploy analogies inspired by the various inanimate machines that were popular in his day—clocks, *tableaux mouvants*, and automata—as to borrow from the animal-based operational metaphors circulating in contemporary medical and natural-philosophical discourse.[42] Nowhere, perhaps, is this mixture of living and artificial machine models more striking than in his depictions of thinkers lost in thought.

Diderot was particularly intrigued by the strange state of sensory oblivion that seemed to occur when a person lost consciousness of everything beyond a single absorbing idea. He found this condition aesthetically appealing: as Michael Fried puts it, it was an 'extreme instance or limiting case' of the interest in absorptive activities evident in the art criticism produced by Diderot and other mid-century theorists.[43] The same interest is apparent in his literary theory, which invested depictions of characters absorbed in *rêverie* with a special power to interest and touch their readers. His major venture into the novel of sensibility, *La Religieuse* (1770/1780–1782), contains a striking example of this idea: at the moment when the heroine Suzanne Simonin is forced against her will to take monastic vows, she turns into an *automate* out of deep dejection and dread for the existence that awaits her—a tableau designed to elicit horror and pity from the novel's inscribed reader, the

---

[39] See Schaffer 1999; Riskin 2003, 2007; Kang 2011.
[40] Starobinski 1972, 16–17.
[41] See Kaitaro 1997, 137–138.
[42] See Kaitaro 2008; Wolfe 1999; Martine 2005.
[43] Fried 1980, 31.

Marquis de Croismare.[44] However, Diderot also found considerable philosophical appeal in extreme mental absorption, especially when the idea or ideas responsible for triggering the state involved abstract thinking.

Diderot shared the view common among contemporary doctors that studious mental application channelled sensibility toward the brain, with immediate and inevitable repercussions for the rest of the body. This is one of the many curious phenomena discussed in *Le Rêve de d'Alembert*, a trio of dialogues whose cast of characters includes a medical authority loosely based on the real-life Théophile de Bordeu. As the fictional Dr. Bordeu explains to the fictional Mlle de l'Espinasse, intense mental exertion concentrates the thinker's vital energy so fully in a single point that it wipes out sensorial awareness of anything else.[45] When translated into the terms of the dialogue's main heuristic metaphor, which compares the brain vis-à-vis the nervous system to a spider at the centre of a web, the meditator's mind becomes the equivalent of a 'spider' that monopolises the organism's vital powers and robs the other parts (the threads) of feeling. Dr. Bordeu describes this shutting out of sensations as a case of the system working backwards, comparable to what happens in delirious fanatics, ecstatic savages, and madmen.[46] Yet he also points out that the phenomenon is not without its advantages: some savvy scholars concentrate their minds on a difficult question as a means of blocking out bodily pain like chronic earache.[47] Such voluntary suppression of physical sensation is, however, only temporary, and the *philosophe* in Dr. Bordeu's case ends up paying with horrible pain for the trick he had tried to play on his sensory system.[48]

More typical, Dr. Bordeu emphasises, is the involuntary oblivion to which cerebralists of the highest order are susceptible. He mentions it in response to a question raised by his second interlocutor, the geometer D'Alembert, who awakes midway through the central dialogue from an agitated, vocal dream. When D'Alembert asks Dr. Bordeu to explain the difference between free will in a dreamer versus a man awake, Dr. Bordeu exclaims:

> You of all people ask me this question! You are a fellow much given to deep speculation, and you have spent two-thirds of your life dreaming with your eyes wide open. In that state, you do all sorts of involuntary things—yes, involuntary—much less deliberately than when you are asleep.[49]

---

[44] '*Je n'entendis rien de ce qu'on disait autour de moi; j'étais presque réduite à l'état d'automate; je ne m'aperçus de rien; j'avais seulement par intervalles comme de petits mouvements convulsifs. On me disait ce qu'il fallait faire; on était souvent obligé de me le répéter, car je n'entendais pas de la première fois, et je le faisais; ce n'était pas que je pensasse à autre c'est que j'étais absorbée; j'avais la tête lasse comme quand on s'est excédé de réflexions [...] On disposa de moi pendant toute cette matinée qui a été nulle dans ma vie, car je n'en ai jamais connu la durée; je ne sais ni ce que j'ai fait, ni ce que j'ai dit*'. Diderot 1770/1975–, 123–24.

[45] Diderot 1769/1975–, 157.

[46] Diderot 1769/1975–, 171.

[47] Diderot 1769/1975–, 173.

[48] Diderot 1769/1975–, 174.

[49] Diderot 1964, 160. Also Diderot 1769/1975–, 184–185.

Dr. Bordeu underscores the odd detachment that occurs between the will and consciousness in both states: whereas the D'Alembert engaged in mathematical speculation might appear to be acting wilfully, he is no more aware of his body's actions than when he is dreaming. As attested by Mlle de l'Espinasse's transcription of his mutterings earlier in the text, D'Alembert carried out an impressive number of seemingly wilful acts while deep in sleep. Yet wilful consciousness was absent from those acts—just as it is, Bordeu insists, when D'Alembert's mind is buried in complex calculations:

> In the midst of your meditations, your eyes are scarcely open in the morning before you are deep in the idea that was on your mind the previous evening. You get dressed, you sit down at the table, you keep on meditating, tracing figures on the cloth; all day long you pursue your calculations; you sit down to dinner; afterwards you pick up your combinations again; sometimes you even get up and leave the table to verify them. You speak with other people, you give orders to your servants, you have a bite of super, you go to bed and you drop off to sleep without having done a single act of your own free will the whole livelong day.[50]

Aram Vartanian has argued that there is something anomalous about this robot-like representation of the geometer actively engaged in thinking, given that 'the ability to think mathematically is anything but automatic'.[51] He contends that the effect of the passage—conveyed rhetorically through its 'lulling', repetitive structure—is to 'defeat our expectation of interiority'.[52] That, however, depends on what expectation we bring to the text. The interiority that Diderot describes here is not psychological: it is organic, molecular even, and one of the most insistent themes of the dialogue is that nothing entirely transcends this level of existence—not individual consciousness, nor the self, nor even God. The character D'Alembert does, indeed, behave rather like an automaton in Dr. Bordeu's portrait of him in wakeful intellectual *rêverie*. In fact, the automaton analogy is even more pronounced in the version of the same anecdote that appears in the *Eléments de physiologie*, where Diderot compares the lack of free will in a geometer preoccupied with a math problem to that of 'a wooden automaton, who carried out the same things as he did'.[53] However, the absorbed geometer is fully interior in the terms that Mlle de l'Espinasse uses elsewhere in the *Rêve* to describe how sensibility is condensed when her mind is fully absorbed by an idea:

> I seem to be reduced to a single point in space; my body almost seems insubstantial, and I am aware only of my thoughts. I am unconscious of location, movement, solidity, distance and space. The universe is annihilated as far as I am concerned, and I am nothing in relation to it.[54]

---

[50] Diderot 1964, 160. ('*Dans le cours de vos méditations, à peine vos yeux s'ouvraient le matin que, ressaisi de l'idée qui vous avait occupé la veille, vous vous vêtiez, vous vous asseyiez à votre table, vous méditiez, vous traciez des figures, vous suiviez des calculs, vous dîniez, vous repreniez vos combinaisons, quelquefois vous quittiez la table pour les vérifier; vous parliez à d'autres, vous donniez des ordres à votre domestique, vous soupiez, vous vous couchiez, vous vous endormiez sans avoir fait le moindre acte de volonté*'. Diderot 1769/1975–, 185.)

[51] Vartanian 1981, 385.

[52] Vartanian 1981, 384, 387.

[53] '*un automate de bois, qui aurait executé les mêmes choses que lui*'. Diderot 1778/1975–, 485–486. Vartanian also discusses this example of the 'conscious automaton' in Vartanian 1981, 382–383.

[54] Diderot 1964, 139.

Like Mlle de l'Espinasse—or the absent-minded Archimedes—the meditating D'Alembert is suspended in space and time, focused so entirely on a particular thought or problem that he can feel nothing else.

Passages like this are, of course, meant to disconcert, and the characters of the *Rêve de d'Alembert* express some fears regarding the loss of wilful, conscious thinking and feeling. The central dialogue begins, we will remember, with Mlle de l'Espinasse explaining that she has called Dr. Bordeu to D'Alembert's bedside because she was alarmed and worried by the strange, disconnected ideas that he was uttering in his sleep. However, as Kate Tunstall stresses, Diderot very deliberately 'refuses any sense of interiority [in the psychological sense] by having D'Alembert's body also express his ideas'—as when D'Alembert masturbates in his sleep after thinking about different forms of possible human generation, thus externalising in a sexual way the ideas that are agitating his mind.[55] Moreover, even though Dr. Bordeu injects the occasional note of pathos into the anecdotes he relates about individuals who, through illness or injury, lose the unified sensibility necessary to have an enduring and coherent sense of self, he and Mlle de l'Espinasse are positively gleeful in the anatomical thought experiments which they conduct in order to carry out that loss in their imagination—as, for example, when they envision reducing the great genius Newton to an 'unorganised pulp' deprived of everything but vitality and sensibility.[56]

In short, the *Rêve de d'Alembert* pushes us, like its fictional interlocutors, to take an externalist perspective and consider deep thinkers (along with everyone else) as living machines with integrated but detachable parts. Viewed from that perspective, consciousness and the other higher faculties of the mind are materially rooted, contingent phenomena whose organic component shows most clearly when those faculties are shut down. In that sense, the absorbed geometer character serves to demonstrate both sides of the 'mechanical' comparison that Diderot made in his *Encyclopédie* article 'Animal'.[57] The geometer's body is just as mechanical in the meditating state as it is in the dream state: what differentiates the two is that the body is more active and efficient under conditions of wakeful mental absorption. To refer one more time to the details of the geometer's day vignette in the *Rêve*: even when D'Alembert's mind is completely wrapped up in a math problem, other parts of his organism—his arms, legs, and stomach—get him up, dressed, fed, and finally back to bed at the end of the day; and as the repeated use of the French imperfect past tense in the passage underscores,[58] they have done so habitually.

The body parts of the conscious automaton in the *Rêve de d'Alembert* thus demonstrate their own particular 'life', a local sensitivity, appetite, and judgement—just

---

[55] Tunstall 2011, 147, 150.

[56] Diderot 1964, 189. On Diderot's materialism, particularly his use of anatomical figures and 'speculative scalpels', see Jacot Grapa 2009, 205–266.

[57] '*Je ne connois rien d'aussi machinal que l'homme absorbé dans une méditation profonde, si ce n'est l'homme plongé dans un profond sommeil*'. (Diderot and Daubenton 1751, 471.) The comparison is designed mainly to undermine Buffon's contention in his *Histoire naturelle* that the human mind always acts voluntarily. See Ann Thomson's analysis of this article in Thomson 1999.

[58] Diderot 1964, 185.

like the eye that, in the *Eléments de physiologie*, helpfully guides an absentminded 'nous' through the streets of Paris:

> How is it that we manage to cross Paris through all sorts of obstacles, when we're deeply preoccupied by an idea? [...] The eye guides us; we're the blind man. The eye is the dog that guides us; and if the eye weren't really an animal reacting to the diversity of sensations, how would it guide us? For this isn't a matter of habit. The obstacles it avoids are at every moment new to it. The eye sees, the eye lives, the eye feels, the eye guides us on, the eye avoids the obstacles, the eye guides us, and guides us surely [...]. The eye is an animal within an animal, carrying out its functions very well, and on its own. The same is true of other organs.[59]

For Diderot, therefore, consciousness is a fleeting, unreliable state, most particularly in those who are prone to get lost in thought. However, the body and its assorted parts keep the whole machine ticking along like clockwork—or, more precisely, like a well-integrated animal economy whose internal parts have their own sort of awareness or attentiveness to their surroundings, along with a capacity for discernment that ensures both the self-preservation and the preservation of the whole. The central focus of all of these vignettes is not the intellectual combinations formed in the absorbed thinker's mind, about which Diderot provides scant information. What he dwells on instead are the operations taking place elsewhere in the thinker's body, operations that he insists here are not purely 'a matter of habit'. By shifting emphasis away from the mind proper, Diderot draws our attention to the dynamic powers that are activated at the organic level when the mind is too busy to notice.[60]

7.4 Conclusion

Over the past two decades, intellectual historians have ventured various interpretations of the place of sensibility in Diderot's model of the mind's operations. Some, like Jonathan Crary, have characterised Diderot as less of a materialist than a mentalist, because he conceived of the senses 'more as adjuncts of a rational mind and less as physiological organs'.[61] Using as his example the blind mathematician Saunderson of the *Lettre sur les aveugles* (1749), he argues that Diderot operated

---

[59] '*Comment se fait-il que nous traversions Paris à travers toutes sortes d'obstacles, profondement occupés d'une idée? [...] L'œil nous mene; nous sommes l'aveugle. L'œil est le chien qui nous conduit; et si l'œil n'était pas reellement un animal se prêtant à la diversité des sensations, comment nous conduirait-il? Car ce n'est pas ici une affaire d'habitude. Les obstacles qu'il evite, sont à chaque instant nouveaux pour lui. L'œil voit, l'œil vit, l'œil sent, l'œil conduit de lui même, l'œil evite les obstacles, l'œil nous mene, et nous mene surement: l'œil ne se trompe que sur les choses qu'il ne voit pas. L'œil est frappé subitement, et il arrête: l'œil accelere, retarde, detourne, veille à sa conservation propre, et à celle du reste de l'equipage; que fait de plus, et de mieux un cocher sur son siege? C'est que l'œil est un animal dans un animal exerçant très bien ses fonctions tout seul. Ainsi des autres organes.*' Diderot 1778/1975–, 499–500.

[60] On Diderot's notions of corporeal memory and corporeal eloquence, see Roach 1993, 116–159.

[61] Crary 1990, 60.

within an epistemological field in which the 'immediate subjective evidence of the body' was less important than the way in which the mind combines the ideas it receives from whatever sensory organs its possesses.[62] Others, like Jessica Riskin and Stephen Gaukroger, have described Diderot as a sensationist with a pronounced moralistic bent. They, too, cite Saunderson but focus less on that character's compensatory ability to think through his fingers than on the cognitive and moral limits created by his 'deficient' physical sensibility, most particularly his 'abstract, inward focus' and lack of compassion for suffering that he cannot see.[63] These traits, they argue, suggest that Diderot was deeply suspicious toward 'solipsistic rationalists' who failed to develop the capacity to participate thoughtfully and compassionately in civic life.

Obviously, different aspects of Diderot's philosophy are at issue in these different interpretations. Crary is interested in Diderot's epistemology inasmuch as it pertains to perception; Gaukroger is preoccupied mainly with Diderot's moral philosophy and psychology; and Riskin is intent on aligning Diderot with other so-called sentimental empiricists, who in her view, placed a premium on both sensory receptiveness and 'emotional and moral openness' to the world.[64] All, however, tend to pass quickly over the materialist, embodied side of Diderot's model of thinking as a function of sensibility. This leads them to miss some of the most intriguing elements of that model.

First, while it is true that Diderot often took a mentalist stance toward the phenomena of sight, touch, and language, his vitalist materialism also led him to espouse a pan-corporeal view of the thinking process. For him, thinking involved the entire animal economy: just as it was perfectly conceivable to imagine that consciousness (the part of the human being that really did the 'thinking') could be located in the fingertips rather than in the head, so, too, it was important to give the inner body its due in the actions and reactions that were unleashed by the process. This perspective was also apparent in his aesthetic theory, where he ascribed a central, sometimes decisive role to the visceral level of human experience in the creation and reception of art.[65] Second, the dominant tone of Diderot's reflections on the occasionally 'cataleptic' behaviour of abstract thinkers was curiosity rather than worry or disapproval: this behaviour interested him because it allowed him to imagine what was going on within the body while the mind was absorbed. When he portrayed absorbed geometers as automatons in *Le Rêve de d'Alembert* and the *Eléments de physiologie*, he used them in much the same way as he did Saunderson: not to imply that they were morally flawed, but, rather, to project himself and his readers into an unfamiliar regimen of sensing and feeling, in order to grasp the

---

[62] Crary 1990, 60.
[63] Riskin 2002, 21–22; Gaukroger 2010, 416.
[64] Riskin 2002, 21.
[65] On the role of the body in Diderot's conception of painting and spectator response, see Brewer 1993, 150–155.

hidden infrastructures of the embodied mind.[66] In other words, as Timo Kaitaro emphasises, methodology was the driving force behind Diderot's life-long interest in people with unusual sensory makeups—a point that Kaitaro makes about Diderot's treatment of those with congenital sensory handicaps, but that can also be applied to his reflections on abstract thinkers.[67]

Finally, Diderot was an anti-reductionist who, particularly in his later years, resisted the temptation to extrapolate moral or psychological truths out of physical sensibility—or vice versa. This is clear both in the 1782 'Additions à la *Lettre sur les aveugles*', where Diderot retracted the claim he made decades earlier that the blind lacked compassion, and in the very structure of the *Rêve de d'Alembert*, where moral issues are largely separated from the conversation on sensibility's physiological and epistemological operations. Sustained ethical discussion is postponed until the third dialogue, where Mlle de l'Espinasse and an increasingly flustered Dr. Bordeu exchange ideas on everything from medically assisted premarital sex to human-animal cross breeding. This tendency to cordon off issues related to moral sensibility when exploring physical sensibility points to a larger tendency of the French Enlightenment: although sensibility was held to be paramount in many realms of human existence—cognitive, affective, social, aesthetic, and physiological—its meanings and operations in those various realms were not conflated into a single moral model. It is worth recalling here the *Encyclopédie*'s definition of sensibility as, first and foremost, a property that heightened or concentrated feeling: that definition left open the possibility that sensibility could have negative moral effects (as it clearly did in Sade) as well as the morally edifying qualities which proponents of sentimentalism ascribed to it.

When we take a broader view of Diderot's depictions of genius types, including those whom he imagined in the state of intellectual absorption, it is clear that his interest in *penseurs profonds* was not rooted in the sentimental moral philosophy he espoused elsewhere (like his 'Eloge de Richardson' and early writings on theatre). He relegated moral sensibility to a decidedly secondary status in some of his best known portrayals of 'creative' absorption, those found in the *Neveu de Rameau* and the *Paradoxe sur le comédien*. Moreover, despite his frequent borrowings from medical discourse, he did not share the concern evident among some contemporary physicians with waking up the senses when they closed as the result of deep thinking. In fact, he insisted that he had done some of his own best deep thinking when he plugged up his ears (as in the *Lettre sur les sourds et muets*), lingered in a dream state (as in the *Salon de 1767*), or cloistered himself for days in his study.

Ultimately, Diderot considered mental absorption and mind-wandering to be productive states that allowed the creative mind to make the complex, unexpected, perhaps aberrant connections among ideas that led to the discovery of truth and beauty. Distraction, as he put it in the *Encyclopédie,* was rooted in 'an excellent

---

[66] As Jacques Chouillet has put it, Diderot undertook in the *Lettre sur les aveugles* to '*explorer avec l'aveugle les ténèbres du monde intérieur, saisir à tâtons les infrastructures et les itinéraires de la compensation, toucher du doigt, s'il se peut, la réalité de l'esprit*'. (Chouillet 1973, 141.)

[67] Kaitaro 1997, 39–50.

quality of the understanding, by which one idea easily sparks another'; and although he cautioned that those capable of being productively distracted should take care not to lose all regard for the people and things around them, he also maintained that 'a good mind must be capable of distractions'.[68] Equally crucial to Diderot's conception of knowledge-seeking was the notion that the flights of genius involve a felicitous alienation, a separation of the conscious mind from its bodily trappings. For the most part, he took a benign view of this sort of alienation, which he also called 'enthusiasm': his genius characters like Dorval of the *Entretiens sur le Fils naturel* do lapse into trance-like oblivion while pondering some aspect of art or nature; however, the condition is temporary and promptly followed by an outpouring of new, inspiring ideas.[69]

Finally, although he shared his century's general veneration for 'deep' thinkers, Diderot did not pathologise them: rather, they have the same tamed quality that Marie-Hélène Huet has noted in his monsters, who are 'safely included in the great chain of being' and fully explainable in material, physical terms.[70] He explicitly rejected the received idea that the scholarly temperament was innately melancholic, and he did not fret over the occasional solipsistic behaviour that he (like many contemporaries) perceived as typical among true intellectuals.[71] If anything, the intellectual personae Diderot invented in his works were blissful in their oblivion to the everyday concerns that agitated lesser minds, like sex and money: they were driven by other passions like glory, renown, and the exquisite pleasure of creating the calculus or a breath-taking work of art. The brain-centred sensibility that they embodied was not sympathetic or socially directed, but that is precisely why it interested him so much.

It is therefore not surprising that Diderot himself may have been absentminded to the point of somnambulism—or at least, so claimed the Montpellier physician Joseph Grasset in his 1907 study *Demifous et demiresponsables*.[72] Although Grasset's claim may be apocryphal, Diderot's correspondence offers some support for the idea that he was deeply drawn toward studious retreat, perhaps to achieve his own moments of productive oblivion. As he wrote to his mistress Sophie Volland in the autumn of 1765, 'My taste for solitude increases by the moment; yesterday, I went out in my dressing gown and nightcap to go dine at d'Amilaville's house. I've taken an aversion to dress clothes; my beard grows as much as it likes'.[73]

---

[68] 'une excellente qualité de l'entendement, une extrême facilité dans les idées de se réveiller les unes les autres'; 'un bon esprit doit être capable de distractions'. Diderot 1754, 1061. On the positive cast given to mind-wandering elsewhere in the *Encyclopédie*, see Bates 2002, 19–40.

[69] As Jean Starobinski argues, Diderot described poetic delirium as a sort of 'fermentation' in his *Encyclopédie* article 'Théosophes'. (Starobinski 1999, 75–80.)

[70] Huet 1993, 89.

[71] '*La mélancolie est une habitude de tempérament avec laquelle on naît et que l'étude ne donne pas. Si l'étude la donnait, tous les hommes studieux en seraient attaqués, ce qui n'est pas vrai*'. Diderot 1875/1975–, 605.

[72] Grasset 1907, 164.

[73] '*Il y aura demain huit jours que je ne suis sorti du cabinet [...] j'ai pris un goût si vif pour l'étude, l'application, et la vie avec moi-même, que je ne suis pas loin du projet de m'y tenir*' and

Diderot's self-description in these letters bears a striking resemblance to the portrait of the absorbed geometer he would soon sketch in the *Rêve de d'Alembert*, both in its emphasis on 'life with myself' and in the reference to venturing out in nightclothes for a social dinner. However, we should also keep in mind that, like all of Diderot's letters to Sophie, these were intended to charm and amuse a mistress, to whom he also wrote:

> My friend, the truth is that we're not made for reading, meditation, letters, philosophy, or sedentary life. It's a depravation for which we pay with our health [...] We shouldn't break altogether with the animal condition, especially since it offers both an infinite number of healthy occupations and some that are quite pleasant, and if I wasn't afraid of scandalising Urania, I'd tell you frankly that I would be healthier if I had spent some of the time I've stayed hunched over my books spread out instead over a woman.[74]

However strongly Diderot may have yearned occasionally to tune out the world and give his mind free reign to chase after ideas, he also felt the pull of the 'animal condition' with all of its needs, tastes, and sometimes extraordinary powers.

References

Alberti, Fay Bound. 2009. Bodies, hearts, minds: Why emotions matter to historians of science and medicine. *Isis* 100: 798–810.

Algazi, Gadi. 2003. Scholars in households: Refiguring the learned habitus, 1480–1550. *Science in Context* 16(1–2): 9–42.

Bates, David W. 2002. *Enlightenment aberrations: Error and revolution in France*. Ithaca/London: Cornell University Press.

Bell, David A. 2001. *The cult of the nation: Inventing nationalism, 1680–1800*. Cambridge, MA: Harvard University Press.

Bonnet, Jean-Claude. 1998. *Naissance du Panthéon: Essai sur le culte des grands hommes*. Paris: Fayard.

Brewer, Daniel. 1993. *The discourse of enlightenment in eighteenth-century France: Diderot and the art of philosophizing*. Cambridge: Cambridge University Press.

Brewer, Daniel. 2008. *The enlightenment past: Reconstructing eighteenth-century French thought*. Cambridge: Cambridge University Press.

Chouillet, Jacques. 1973. *La formation des idées esthétiques de Diderot, 1745–1763*. Paris: Colin.

Condillac, Etienne Bonnot de. 1746/1973. *Essai sur l'origine des connaissances humaines*. Paris: Galilée.

'*Mon goût pour la solitude s'accroît de moment en moment; hier je sortis en robe de chambre et en bonnet de nuit, pour aller dîner chez d'Amilaville. J'ai pris en aversion l'habit de visite; ma barbe croît tant qu'il lui plaît*'. Diderot 1997, 541–542, 556.

[74] '*Tenez, mon amie, c'est que nous ne sommes pas destinés à la lecture, à la méditation, aux lettres, à la philosophie, et à la vie sédentaire [...] Il ne faut pas rompre tout à fait avec la condition animale; d'autant que cette condition, parmie une infinité d'occupations saines, en offre plusieurs qui sont assez plaisantes, et si je ne craignais de scandaliser Uranie, je vous dirais franchement que je me porterais mieux si j'étais resté penché sur une femme une portion du temps que je suis resté penché sur mes livres*'. Diderot, letter to Sophie Volland, 7 November 1762, in Diderot 1997, 470.

Condillac, Etienne Bonnot de. 1754/1984. *Traité des sensations et Traité des animaux*. Paris: Fayard.
Condren, Conal, Stephen Gaukroger, and Ian Hunter (eds.). 2006. *The philosopher in early modern Europe: The nature of a contested identity*. Cambridge: Cambridge University Press.
Cook, Harold J. 2002. Body and passions: Materialism and the early modern state. *Osiris* 2nd series, 17: 25–48.
Cook, Alexander. 2009. The politics of pleasure talk in eighteenth-century Europe. *Sexualities* 12: 451–466.
Crary, Jonathan. 1990. *Techniques of the observer: On vision and modernity in the nineteenth century*. Cambridge, MA: MIT Press.
Daston, Lorraine. 2001. Enlightenment fears, fears of enlightenment. In *What's left of enlightenment? A post-modern question*, ed. Keith M. Baker and Peter H. Reill, 115–128. Stanford: Stanford University Press.
Daston, Lorraine. 2004. Attention and the values of nature. In *The moral authority of nature*, ed. Lorraine Daston and Fernando Vidal, 100–126. Chicago: University of Chicago Press.
Diderot, Denis.1754. Distraction. In *Encyclopédie ou Dictionnaire raisonné des arts et des métiers*, 35 vols., vol. 4, ed. Denis Diderot and Jean Le Rond D'Alembert, 1061. Paris: Briasson, David, Le Breton & Durand.
Diderot, Denis. 1769/1975–. Le Rêve de d'Alembert. In *Oeuvres complètes*, 25 vols., vol. 17, ed. Herbert Dieckmann, Jacques Proust, Jean Varloot et al., 25–209. Paris: Hermann.
Diderot, Denis. 1770/1975–. La Religieuse. In *Oeuvres complètes*, 25 vols., vol. 11, ed. Herbert Dieckmann, Jacques Proust, Jean Varloot, et al., 3–302. Paris: Hermann.
Diderot, Denis. 1778/1975–. Eléments de physiologie. In *Oeuvres complètes*, 25 vols., vol. 17, ed. Herbert Dieckmann, Jacques Proust, Jean Varloot et al., 261–574. Paris: Hermann.
Diderot, Denis. 1875/1975–. Réfutation suivie de l'ouvrage d'Helvétius intitulé *l'Homme*. In *Oeuvres complètes*, 25 vols., vol. 24, ed. Herbert Dieckmann, Jacques Proust, Jean Varloot et al., 423–767. Paris: Hermann.
Diderot, Denis. 1964. D'Alembert dream. In *Rameau's Nephew and Other Works*. Trans. Jacques Barzun and Ralph Bowen, 92–175. Indianapolis: Bobbs-Merrill.
Diderot, Denis. 1997. *Correspondance*, Oeuvres, vol. 5. Paris: R. Laffont.
Diderot, Denis, and Louis Jean-Marie Daubenton. 1751. Animal. In *Encyclopédie ou Dictionnaire raisonné des arts et des métiers*, 35 vols., vol. 1, ed. Denis Diderot and Jean Le Rond D'Alembert, 468–474. Paris: Briasson, David, Le Breton & Durand.
Figlio, Karl. 1975. Theories of perception and the physiology of mind in the late eighteenth century. *History of Science* 13: 177–212.
Fouquet, Henri. 1765. Sensibilité, Sentiment (Médecine). In *Encyclopédie ou Dictionnaire raisonné des arts et des métiers*, 35 vols., vol. 15, ed. Denis Diderot and Jean Le Rond D'Alembert, 38–52. Paris: Briasson, David, Le Breton & Durand.
Fried, Michael. 1980. *Absorption and theatricality: Painting and beholder in the age of Diderot*. Chicago: University of Chicago Press.
Gaukroger, Stephen. 2010. *The collapse of mechanism and the rise of sensibility: Science and the shaping of modernity, 1680–1760*. Oxford: Oxford University Press.
Grasset, Joseph. 1907. *Demifous et demiresponsables*. Paris: Alcan.
Huet, Marie-Hélène. 1993. *Monstrous imagination*. Cambridge, MA: Harvard University Press.
Jacot Grapa, Caroline. 2009. *Dans le vif du sujet: Diderot corps et âme*. Paris: Editions Classiques Garnier.
Janković, Vladimir. 2010. *Confronting the climate: British Airs and the making of environmental medicine*. Basingstoke: Palgrave Macmillan.
Jaucourt, Louis de. 1756. Etude. In *Encyclopédie ou Dictionnaire raisonné des arts et des métiers*, 35 vols., vol. 6, ed. Denis Diderot and Jean Le Rond D'Alembert, 86–96. Paris: Briasson, David, Le Breton & Durand.
Jaucourt, Louis de. 1765. Sensibilité (Morale). In *Encyclopédie ou Dictionnaire raisonné des arts et des métiers*, 35 vols., vol. 15, ed. Denis Diderot and Jean Le Rond D'Alembert, 52. Paris: Briasson, David, Le Breton & Durand.

Kaitaro, Timo. 1997. *Diderot's Holism: Philosophical anti-reductionism and its medical background.* Frankfurt: P. Lang.
Kaitaro, Timo. 2007. Emotional pathologies and reason in French medical enlightenment. In *Forming the mind: Essays on the internal senses and the mind/body problem from Avicenna to the medical enlightenment,* ed. Henrik Lagerlund, 311–325. Dordrecht: Springer.
Kaitaro, Timo. 2008. Can matter mark the hours? Eighteenth-century vitalist materialism and functional properties. *Science in Context* 21(4): 581–592.
Kang, Minsoo. 2011. *Sublime dreams of living machines: The automaton in the European imagination.* Cambridge, MA: Harvard University Press.
Martine, Jean-Luc. 2005. L'article ART de Diderot: Machine et pensée pratique. *Recherches sur Diderot et sur l'Encyclopédie* 39: 2–29.
Mercier, Louis-Sébastian. 1764/1776. Discours sur la lecture. In *Eloges et discours philosophiques,* 233–296. Amsterdam: E. van Harrevelt.
Mercier, Louis-Sébastian. 1766/1776. Le Bonheur des gens de lettres. In *Eloges et discours philosophiques,* 3–58. Amsterdam: E. van Harrevelt.
Moreau, Pierre. 1959–1960. Fontenelle. In *Le Dix-huitième siècle,* 2 vols., vol. 1, ed. Georges Grente et al., 460–466. Dictionnaire des lettres françaises, vol. 4. Paris: A. Fayard.
Plutarch. 1683–1686/2008. Marcellus. In *Plutarch's Lives of Illustrious Men.* Trans. John Dryden, 3 vols., vol. 1, 470–497. Wildside Press
Rey, Roselyne. 2000. *Naissance et développement du vitalisme en France de la deuxième moitié du 18e siècle à la fin du Premier Empire.* Oxford: Voltaire Foundation.
Ribard, Dinah. 2003. *Raconter, vivre, penser: histoires de philosophes, 1650–1766.* Paris: Vrin.
Riskin, Jessica. 2002. *Science in the age of sensibility: The sentimental empiricists of the French enlightenment.* Chicago: University of Chicago Press.
Riskin, Jessica. 2003. The defecating duck, or, the ambiguous origins of artificial life. *Critical Inquiry* 29(4): 599–633.
Riskin, Jessica (ed.). 2007. *Genesis redux: Essays in the history and philosophy of artificial life.* Chicago: University of Chicago Press.
Roach, Joseph. 1993. *The player's passion: Studies in the science of acting.* Ann Arbor: University of Michigan Press.
Schaffer, Simon. 1999. Enlightened automata. In *The sciences in enlightened Europe,* ed. William Clark, Jan Golinski, and Simon Schaffer, 126–166. Chicago: University of Chicago Press.
Starobinski, Jean. 1972. Diderot et la parole des autres. *Critique* 296: 3–22.
Starobinski, Jean. 1999. *Action et réaction: vie et aventures d'un couple.* Paris: Seuil.
Steinke, Hubert. 2005. *Irritating experiments: Haller's concept and the European controversy on irritability and sensibility, 1750–1790.* Amsterdam/New York: Rodopi.
Sutton, John. 2010. Carelessness and inattention: Mind-wandering and the physiology of fantasy from Locke to Hume. In *The body as object and instrument of knowledge: Embodied empiricism in early modern science,* ed. Charles T. Wolfe and Ofer Gal, 243–263. Dordrecht/Heidelberg/London/New York: Springer.
Thomson, Ann. 1999. Diderot, le matérialisme et la division de l'espèce humaine. *Recherches sur Diderot et sur l'Encyclopédie* 26: 197–211.
Tissot, Samuel Auguste André David. 1769. *An essay on diseases incident to literary and sedentary persons, with proper rules for preventing their fatal consequences, and instructions for their cure, with a preface and notes by J. Kirkpatrick, M.D.* London: Norse and Dilly.
Tissot, Samuel Auguste André David. 1775. *De la Santé des gens de lettres,* 3rd ed. augmentée. Lausanne: Chez François Grasset.
Tunstall, Kate E. 2011. Eyes wide shut: *Le Rêve de d'Alembert.* In *New essays on Diderot,* ed. James Fowler, 141–157. Cambridge: Cambridge University Press.
Vandermonde, Charles Augustin. 1759/1760. *Dictionnaire portatif de santé,* nouvelle ed. 2 vols. Paris: Chez Vincent.
Vartanian, Aram. 1981. Diderot's Rhetoric of Paradox, or, the conscious automaton observed. *Eighteenth-Century Studies* 14(4): 379–405.

Venel, Gabriel. 1754. Digestion. In *Encyclopédie ou Dictionnaire raisonné des arts et des métiers*, 35 vols., vol. 4, ed. Denis Diderot and Jean Le Rond D'Alembert, 999–1003. Paris: Briasson, David, Le Breton & Durand.
Vidal, Fernando. 2009. Brainhood, anthropological figure of modernity. *History of the Human Sciences* 22(5): 5–36.
Vila, Anne C. 1998. *Enlightenment and pathology. Sensibility in the literature and medicine of eighteenth-century France*. Baltimore: Johns Hopkins University Press.
Williams, Elizabeth A. 1994. *The physical and the moral: Anthropology, physiology and philosophical medicine in France, 1750–1850*. Cambridge: Cambridge University Press.
Williams, Elizabeth A. 2003. *A cultural history of medical vitalism in enlightenment Montpellier*. Aldershot: Ashgate.
Wilson, Catherine. 2008. The enlightenment philosopher as social critic. *Intellectual History Review* 18(3): 413–425.
Wolfe, Charles. 1999. Machine et organisme chez Diderot. *Recherches sur Diderot et sur l'Encyclopédie* 26: 213–231.
Wolfe, Charles T., and Motoichi Terada. 2008. The animal economy as object and program in Montpellier vitalism. *Science in Context* 21(4): 537–579.
Yvon, Claude, and Jean-Henri-Samuel Formey. 1751. Attention. In *Encyclopédie ou Dictionnaire raisonné des arts et des métiers*, 35 vols., vol. 1, ed. Denis Diderot and Jean Le Rond D'Alembert, 840–843. Paris: Briasson, David, Le Breton & Durand.
Zimmermann, Johann-Georg. 1774/1855. *Traité de l'expérience en général, et en particulier dans l'art de guérir*. Trans. Jean-Baptiste Lefebvre de Villebrune, 3 vols. Paris: chez Vincent. Original German edition: Zimmermann, Johann-Georg. 1763–1764. *Von der Erfahrung in der Arzneikunst*. Zurich: Heidegger und Compagnie.

Chapter 8
Sensibility as Vital Force or as Property of Matter in Mid-Eighteenth-Century Debates

Charles T. Wolfe

Abstract Sensibility, whether understood in moral, physical, medical, or aesthetic terms, seems to be a paramount case of a *higher-level*, intentional property, not a *basic* property. Diderot famously claimed that *matter itself* senses, with sensibility being a general or universal property of matter, even if he sometimes stepped back from this claim and called it a 'supposition'. Crucially, sensibility here is a 'booster': it enables materialism to account for the phenomena of conscious, sentient life, contrary to what its opponents hold, for if matter can sense, and sensibility is not merely a mechanical process, then the loftiest cognitive plateaus belong to one and the same world as the rest of matter. Lelarge de Lignac noted this when he criticised Buffon for 'granting to the body [*la machine*, a then-common term for the body] a quality which is essential to minds, namely sensibility'. This view, which Diderot definitely held, was comparatively rare, stemming from medico-physiological sources including Robert Whytt, Albrecht von Haller, and Théophile de Bordeu. We then have, I suggest, an intellectual landscape in which newly articulated properties such as irritability and sensibility are presented either as experimental properties of muscle fibres to be understood mechanistically (Hallerian irritability), or as properties of matter itself (whether specifically *living* matter as in Bordeu and his fellow *montpelliérains* Ménuret and Fouquet, or matter in general, as in Diderot). I am not convinced that their debates involve an identical concept, but nevertheless propose a topography of the problem of sensibility as property of matter or as vital force in mid-eighteenth-century debates—not an exhaustive cartography of all possible theories, but an attempt to understand the 'triangulation' of three views: a vitalist view in which sensibility is fundamental, matching up with a conception of the organism as the sum of parts conceived as little *lives* (Bordeu et al.); a broadly mechanist view which builds upwards, step by step, from the basic property of irritability to the higher-level property of sensibility (Haller); and, more eclectic, a materialist view

---

C.T. Wolfe (✉)
Department of Philosophy and Moral Sciences, Sarton Centre for History of Science,
GhentUniversity, Blandijnberg 2, B-9000 Ghent, Belgium
e-mail: charles.wolfe@ugent.be

which seeks to combine the explanatory force of the Hallerian approach with the metaphysically explosive (monistic) potential of the vitalist approach (Diderot). Examining Diderot in the context of this triangulated topography of sensibility as property should shed light on his famous proclamation regarding sensibility as a universal property of matter.

> *Sensibilité, Sentiment (Médecine): la faculté de sentir, le principe sensitif, ou le sentiment même des parties, la base et l'agent conservateur de la vie, l'animalité par excellence, le plus beau, le plus singulier phénomène de la nature, etc.*[1]

Sensibility, in any of its myriad realms—moral, physical, aesthetic, medical, and so on—seems to be a paramount case of a *higher-level*, intentional property, not a *basic* property. That is, while we sometimes suspect, or at least pretend to suspect that rocks can sense, we do not consider sensibility an 'atomic' property like shape, size, and motion. Higher-level properties like sensibility, thought, memory, desire seem to belong to higher organisms, which leaves room for debate (lizards have recently, as of early 2012, been shown to display learning abilities which lead them to be classified higher up the cognitive scale—and of course the idea of a 'higher organism' is itself a piece of folk biology). Now, materialism is often considered to reduce all higher-level properties of our experience to basic ones such as, precisely, shape, size, and motion—which was of course the program of the mechanical philosophy in the seventeenth century. This leads to the once-frequent view that materialism is necessarily *mechanistic* materialism; as a recent entry in a noted secondary source, the *Oxford Companion to the History of Modern Science*, tells us, 'materialists explain everything in terms of matter and motion; vitalists, in terms of the soul or vital force'.[2] But anyone who reads a page of Diderot, to name one notable example, finds a very different constellation from this commonplace opposition between 'matter' and 'sensibility'.

Diderot famously made the bold and attributive move of postulating that *matter itself* senses, or that sensibility (perhaps better translated 'sensitivity' here, although for the sake of consistency, I will keep the older 'sensibility'[3]) is a general or universal property of matter, even if he at times took a step back from this claim and called it a 'supposition'. Crucially, sensibility is here playing the role of a 'booster': it enables materialism to provide a full and rich account of the phenomena of conscious, sentient life, contrary to what its opponents hold: for if matter can sense, and sensibility is not a merely mechanical process, then the loftiest cognitive plateaus

---

[1] Fouquet 1765, 38b.

[2] Wellman 2003.

[3] In this paper I use the English 'sensibility' for the French *sensibilité*, as it was the common term at the time, but it should be clear that I mean 'sensitivity': the property of organic beings to sense and respond to stimuli or impressions. Thus Haller's classic paper of 1752 (published in an English translation in 1755) is *A treatise on the sensible and irritable parts of animals*, not on their 'sensitive' parts. 'Sensibility' in this context is not, say, a term from moral philosophy but an organic term.

are accessible to materialist analysis, or at least belong to one and the same world as the rest of matter.

This was noted by the astute anti-materialist critic, the Abbé Lelarge de Lignac, who, in his 1751 *Lettres à un Amériquain*, criticised Buffon, the great naturalist, author of the 15-volume *Histoire naturelle* (and its seven-volume *Supplément*), but also theorist of generation, for 'granting to the body [*la machine*, a common term for the body at the time] a quality which is essential to minds, namely sensibility'.[4] This view, here attributed to Buffon and definitely held by Diderot, was comparatively rare. If we look for the sources of this concept, the most notable ones are physiological and medical treatises by prominent figures such as the Edinburgh professor of medicine Robert Whytt (1714–1766), the Swiss, but Göttingen-based Albrecht von Haller (1709–1777), and the Montpellier physician Théophile de Bordeu (1722–1776), the latter being a key representative of the school we customarily refer to as the Montpellier vitalists. We then have, or so I shall try to sketch out, an intellectual landscape in which new—or newly articulated—properties such as irritability and sensibility are presented either as an experimental property of muscle fibres that can be understood mechanistically (Hallerian irritability, as studied recently by Hubert Steinke), or a property of matter itself (whether specifically *living* matter as in Bordeu and his fellow *montpelliérains* Ménuret and Fouquet, or matter in general, as in Diderot).

I am by no means convinced that it is one and the same 'sensibility' that is at issue in debates between these figures (as when Bordeu attacks Haller's distinction between irritability and sensibility and claims that 'his own' property of sensibility is both more correct and more fundamental in organic beings), but I am interested in mapping out a topography of the problem of sensibility as property of matter or as vital force in mid-eighteenth-century debates—not an exhaustive cartography of all possible positions or theories, but an attempt to understand the 'triangulation' of three views: a mechanist, or 'enhanced mechanist' view in which one can work upwards, step by step from the basic property of irritability to the higher-level property of sensibility (Haller); a vitalist view in which sensibility is fundamental, matching up with a conception of the organism as the sum of parts conceived as little *lives* (Bordeu et al.); and, more eclectic, a materialist view which seeks to combine the mechanistic, componential rigour and explanatory power of the Hallerian approach, with the monistic and metaphysically explosive potential of the vitalist approach (Diderot). It is my hope that examining Diderot in the context of this triangulated topography of sensibility as property sheds light on his famous proclamation regarding sensibility as a universal property of matter: 'sensibility is a universal property of matter'.[5]

---

[4] To be precise, Lignac is following Condillac's criticism of Buffon, but he adds that Condillac is just as guilty of error since he 'attributes to the soul that which belongs solely to the machine'. Lelarge de Lignac, quoted by Condillac, *Lettre à l'auteur des Lettres à un Amériquain*, annexed to *Traité des animaux* (1755), in Condillac 1754/1984, 425.

[5] Diderot, letter to Duclos, 10 October 1765, in Diderot 1955–1961, Vol. 5, 141. As I discuss below, he also calls it a 'general property of matter' and in other texts, casts doubt on this hypothesis.

8.1 Irritability/Sensibility as Commodity or Danger: A Hallerian Context

La sensibilité fait le caractère essentiel de l'animal[6]

The idea that certain types of organic matter possess reactive or even reflexive properties which were termed 'irritability' and 'sensibility' was, if not 'in the air' in a vague Zeitgeist-like sense, definitely discussed by a variety of figures across early modern Europe, in differing contexts (more or less experimentalist, more or less 'philosophical', more or less prestigious, and so on). While the history of these debates has largely been mapped out,[7] it is important for my purposes here to provide some reconstruction of this material—not least since it is so difficult to separate 'experimental' work or aspects from 'philosophical' statements or appropriations of something purportedly experimental.

The physician Francis Glisson (1598–1677), great authority on the liver, gall bladder, and rickets (in works such as his 1654 *Anatomia hepatis*), and Regius Professor of Physic at Cambridge, is the locus classicus for the property of irritability—a term which he coined (*irritabilitas*), as Albrecht von Haller noted. After writing a number of such medical treatises, he produced the *Tractatus de natura substantiae energetica, seu de vita naturae* (1672), a metaphysics of living nature in which a rudimentary level of perception was posited as existing in matter itself. Matter contains, he stated, the root of life. Just as particular organs have a capacity to react to certain stimuli, so ultimately did matter itself. Irritability was the equivalent at the functional level to the basic property of 'natural perception' in matter.[8]

Albrecht von Haller's concept of irritability, in contrast, has a distinctly experimental flavour—measuring the reaction of parts of the body that did not seem to transmit their stimulation to the 'soul' (which would be tantamount to reflexivity). This is the basic definition of how irritability differs from sensibility:

> I call that part of the human body irritable, which becomes shorter upon being touched; very irritable if it contracts upon a slight touch, and the contrary if by a violent touch it contracts but little. I call that a sensible part of the human body, which upon being touched transmits the impression of it to the soul.[9]

This force cannot come from the nerves, since even after they have been cut, muscular fibres can still be irritated, and contract.[10] Sensations are caused by

---

[6] Haller 1777, 776a (the first sentence of the article).

[7] For the later debates on irritability and sensibility, see Duchesneau 1982, 1999; Vila 1998; Steinke 2005; for the earlier appearance of the concept of irritability, see Giglioni 2008.

[8] Guido Giglioni's various essays on Glisson are fascinating studies of this figure and broader issues in the history and philosophy of early modern life science. See, most recently, Giglioni 2008, 465–493.

[9] Haller 1755/1936, 4–5.

[10] Haller 1755/1936, 39.

impressions of objects on the nerves that transmit the impetus to the brain, and from there onto the soul.[11]

Irritability is a quantifiable, experimentally accessible property of the muscle fibres, to be studied mechanistically, in the sense that there will be a correlation between a measurable degree of irritation and a degree of irritation of the fibres: between structure and function. There is no metaphysics of living matter here, at least in appearance. For on the one hand, to be sure, Haller wants to define irritability in such a way as to rule out 'speculative hidden qualities'.[12] But on the other hand, when pushed as to the reason why certain types of organic matter possess such properties, Haller first attributes it to the 'gluten' within the fibre ('irritability is actually a force specific to animal gluten',[13] although he wavers on this), and then, coming dangerously close to just as vitalist a metaphysics as Glisson (or just as metaphysical a vitalism), attributes this 'vitality' to a hidden force, the *visinsita*.[14] Sometimes he is more cautious, and either rejects such considerations as overly philosophical (as when he wants to disqualify La Mettrie's radical appropriations of his work, turning irritability into a material basis for life), or plays the agnostic, declaring as regards the ultimate cause of irritability that the alleged source 'lies concealed beyond the reach of the knife and microscope'.[15] Haller the pupil of Boerhaave, the tireless vivisectionist, the inventor of ingenious Newtonian-inspired or otherwise 'geometric' methods and concepts for quantifying the hitherto mysterious properties of life, is himself something of a vital force thinker. Positioning him correctly on an *échiquier des possibles* of eighteenth-century debates combining, as they do, the metaphysics of the soul and the physiology of muscular motion, is easy in some respects, not least given his development of an experimental method and a 'protocol' by which different members of a laboratory can reproduce experiments, but it is difficult when it comes to metaphysical commitments.

For Haller does not want irritability to be presented as a material basis for life in the sense of *materialism* (as is explicit in his polemic with La Mettrie).[16] He wishes

---

[11] Haller 1755/1936, 4.

[12] Steinke 2005, 106.

[13] Haller to Bonnet, 15 March 1755, in Sonntag, ed. 1983, 63.

[14] This intriguing expression does not appear in Haller's early lecture ('Dissertation') on sensibility and irritability, but, as Steinke has noted (Steinke 2005, 106, 123) only in the later *Elementa physiologiae* and revised editions of the *Primae lineae physiologiae*, e.g. 'The heart and intestines, also the organs of generation, are governed by a *vis insita*, and by stimuli. These powers do not arise from the will; nor are they lessened or excited, or suppressed, or changed by the same. No custom no art can make these organs subject to the will, which have their motions from a *vis insita*; nor can it be brought about, that they should obey the commands of the soul, like attendants on voluntary motion' (Haller 1779, Sect. 409, 198–199); the original Latin is in Haller 1747/1765, Chap. 9, Sect. 409, 184. The passage is misattributed in Elizabeth L. Haigh's otherwise excellent study to Haller's earlier Dissertation (Haigh 1984, 52). It is also used, without attribution, in the 'Anatomy' article of the *Encyclopaedia Londinensis*, Wilkes, ed. 1810, 563–564. (Thanks to Trevor Pearce, Lucian Petrescu, and Kimberly Garmoe for help with the correct attribution, as well as to Hubert Steinke for his assistance over time.)

[15] Haller 1755/1936, 8.

[16] See Roe 1984, esp. 282–284.

to preserve an independent 'arena' or space of existence for the soul, which is partly contingent on the distinction between irritability—belonging only to the muscles—and sensibility—which has to 'report' to the soul. This is also part of his disagreement with what I shall call below the 'sensibility monism' of vitalists such as Bordeu—a point further extended against Haller by Paul-Joseph Barthez (1734–1806) and other *montpelliérains*: there is an experimental disagreement, there is a disagreement about the place of philosophical considerations in medical practice,[17] but above all, Haller fears a scenario in which matter itself is alive, whether it is an 'irritable matter'—La Mettrie's—or a 'sensible matter'—Bordeu's, that of other vitalists overall, and Diderot's.

Conversely, Haller also disagrees with Robert Whytt, a professor of medicine at Edinburgh, for giving *too much* room to the soul. Whytt's 1751 work *An Essay on the Vital and Other Involuntary Motions of Animals* provided a general theory of sensibility, which he viewed as *primary* with respect to irritability. Whytt associated sensibility and life under the heading of one 'active sentient principle', which however he insisted could not be a mere property of matter itself.[18] Put differently, irritability presupposes sensibility, so that the latter is not the sole exclusive property of the nerves (which were taken to include, not just the conduit, but the 'nervous substance' itself). Rather, it is distributed throughout the body, whereas for Haller, as we saw, certain organs and tissue types are insensible. Revealingly for our purposes, Haller more than once assimilated Whytt's view to Stahlian animism (the view that all active functions in the body are somehow the doings of the soul, which, despite being immaterial, is nevertheless controlling the body).[19] Rather than a monism of an active sentient principle, a variant (like Glisson's, but differently) of a vision of active matter, Haller promotes a structural model, that is, 'a decentralization of active powers within the animal economy'.[20] A key implication of this decentralised view is that irritability does not have 'anything in common with the soul', as Haller put it.[21] There is both a *functional* reason for this (the distinction between two types of properties but also two *levels*), and a *metaphysical* reason: both Whytt and La Mettrie pose metaphysical dangers, not so much 'animism or materialism' as they are usually presented, but really, materialism *simpliciter*, understood as a theory which explains the higher-level in terms of the lower-level.

If the problem is materialism, then it may even be artificial to separate the issue into *levels*—of matter, of functions, etc.—versus *metaphysics*: for the concern with

---

[17] Boury 2008, 521–535.

[18] Whytt 1768, 128.

[19] The debate (rather acrimonious as it was) continued for years: Whytt replied to Haller in his 'Observations on the Sensibility and Irritability of the Parts of Man and Other Animals: occasioned by Dr. Haller's late Treatise on these Subjects', in Whytt 1768; Haller's later *Mémoires* on sensibility and irritability are, amongst other things, are a further reply to Whytt. For further analysis see French 1969, 9, 63; Duchesneau 1982, Chap. 6; Steinke 2005, Chap. 3.

[20] Reill 2005, 131.

[21] Haller 1756–1760, Vol. 1, 91.

levels is a metaphysical concern, with the lower and the higher. As Roger French comments nicely,

> Haller reserves the adjective 'sensible' for those organs or tissues which are capable of communicating to the soul within the brain and there arousing a conscious sensation. He therefore never accepted Whytt's notion of *unconscious* sensation, a mere lowly animal 'feeling' of the sort that allowed oysters to close up at the approach of danger'.[22]

I hope it is clear that, as in the other episodes of our story, what is at issue is an act of *attribution* of higher-level properties to a lower-level substrate; and more broadly, the articulation of a concept of living matter in which sensibility is the operative property. Haller himself—not Glisson, not Whytt, not Bordeu, and not Diderot—states that '[s]ensibility is the essential trait of the animal. That which senses is an animal, that which does not sense is not'[23] (the latter two thinkers do say such things, but my point is that here it is Haller himself speaking).

The story of irritability and sensibility, and their provenance and derivations in this period could be extended much further (with, e.g., Baglivi, Stahl, and Bonnet) but as I indicated at the outset, my aim is more limited in the sense that I want to contrast three positions: higher-level properties as mechanistically specifiable properties of certain types of matter (Haller), as features of all living, organised animal matter[24]—organised as a system of interconnecting 'little lives' (the vitalist view), and lastly, as universal properties of matter itself (materialism, in its Diderotian variant). What is noteworthy so far is that even in the most mechanism-friendly part of the story, Haller's, the risk of slipping into a form of vitalism (for there are many forms of vitalism![25]) is constant, and perhaps made all the more explicit by the way in which figures like Glisson, and later Whytt or Stahl, need to be portrayed as defending purely idealistic, experimentally unsound or ungrounded metaphysics of life, as distinct from a more naturalistically grounded scientific study of organisms.

If Glisson's approach was an attribution of higher-level properties to a lower level he called 'living nature', which was negatively portrayed by Haller so as to guarantee his own experimental, scientific legitimacy, while presenting his predecessor as a mysterious-force vitalist,[26] the tension between Haller, Bordeu and the vitalists, and Diderot (who is in more of a 'dialectical' position with respect to the others) shows that a linear portrayal of the debate is a hopeless task, particularly a 'positivistic' account in which thinkers gradually move from metaphysical

---

[22] French 1969, 71.
[23] Haller 1777, 776a (the first sentence of the article).
[24] For the case of plants see Garrett 2003, 63–81.
[25] Wolfe 2011b.
[26] Ironically, even Glisson needed to follow this procedure and distinguish his own metaphysics of appetite, perception, and living nature from the views of a *more* monistic, more vitalistic, and thus more radical thinker, the Renaissance naturalist Tommaso Campanella. For Glisson, Campanella 'assigned to inanimate material beings more than I would like to, that is, sensation itself'(Campanella 1672).

speculation to 'real science' via experimental trial and error. That is, as I shall indicate in closing, there is a permanent vitalist remainder in the attribution of a mind-like, reactive, and/or intentional property to a system of organised matter. Not only are the above-mentioned tensions not empirically resolvable (as if it were a matter of deciding between three theories of reflex action, or three disciplinary definitions of the role of physiology); their lack of resolution is also not just ideological (e.g. regarding commitments to a preserved space for the soul, given a naturalistic account of mental life), but metaphysical: the fear of attributing higher-level properties to a basic substrate, such as matter. Curiously, however, there is no neat separation between orthodox dualists and heterodox materialists here. Notably, because all parties, as I have noted, keep on slipping into various kinds of vitalism—never in the sense of mysterious vital forces like Hans Driesch's entelechies in the late nineteenth and early twentieth centuries, but rather in the sense of the insistence on the uniqueness of the functional properties of certain types of material arrangements, namely, arrangements that form 'organised wholes', also known as *corps organisés*, or 'organisms' in our vocabulary, or 'animal economies' to use the period's term. Now, where this slippage into monism (since here the vitalist concept, or family of concepts, is one and the same as the monist concept[27]) frightens some thinkers, it is on the contrary a desirable outcome to others, not least since it allows for a naturalistically respectable way of dealing with complex properties: what I call sensibility as a 'booster' of matter.

8.2 Sensibility as Go-Between or Unifier: Vitalist Scenarios

la doctrine de la sensibilité [est] la même avec celle du *vitalisme*[28]

When we speak of the Montpellier vitalists, we are referring to the group of physicians and professors of medicine (but also anatomy, botany, etc.) at the Faculty of Medicine at Montpellier, beginning in the mid-eighteenth century; the term 'vitalist' was applied to this group from approximately 1800, and indeed served as a self-description during those decades, although some, like Paul-Joseph Barthez, declared, after most of the influential works—by La Caze, Bordeu et al.—had already been published, that he did not 'wish to be the Leader of the Sect of the Vitalists'.[29] Given their shared insistence on sensibility as the sole, defining property of living beings, against Haller's basic distinction between irritability and sensibility, the vitalists could just as easily have been called 'sensibilists'; although in the end, Henri Fouquet, when reflecting retrospectively on their movement in an 1803 work,

---

[27] On sensibility as a monistic property in the 'philosophical medicine' of the *montpelliérains* see Vila 1998, Chap. 2.
[28] Fouquet 1803, 78, n. 5.
[29] Barthez 1806, Vol. 1, 98, N. 18. The first edition appeared in 1778.

simply stated that the terms amount to the same thing, since whatever is sensitive (or sensible) is vital ('everything that senses, is vital'[30]).

With the vitalists, two major transformations occur with regard to the concept of sensibility as we have encountered it, primarily in its Hallerian presentation.

Empirically—or at a level presented as empirical, experimentally founded, observationally documented, and so on—sensibility is now presented as the primary and general property of living beings (tantamount to life, as Fouquet says above), so that the distinction between irritability and sensibility is jettisoned. To take two examples amongst many, Gabriel-François Venel (1723–1775), a chemist and physician who was close to Théophile de Bordeu, and authored the long, dense entry 'Chymie' in the *Encyclopédie*, stated in the two-line entry 'Irritabilité (Physiologie)', which is mainly a *renvoi* to Fouquet's long entry 'Sensibilité', that irritability was a word invented by Glisson, then revived 'nowadays by the famous Mr Haller', 'to refer to a particular mode of a more general faculty of the organic parts of animals, which we will discuss under the name "sensibility"'.[31] Irritability is just a mode of a more general and primary property, sensibility. Another, brilliant and under-studied, Montpellier vitalist figure was Jean-Joseph Ménuret de Chambaud (1733–1815), whom I shall not discuss in detail here.[32] In his fascinating article 'Œconomie Animale', Ménuret, too, refers to the property Glisson called irritability, in order to fold it into the more essential property of sensibility. The basic features of life, Ménuret argues, are 'movement and feeling (*sentiment*)' and these are 'probably reducible to one basic (*primitif*) kind', a yet more basic property, a 'singular property, the source of movement and feeling as connected to the *organic* nature of the elements composing the body'. Ménuret adds that this property depends on a unique type of union between molecules, which Francis Glisson discovered, and named *irritability*—but in fact, it is really just a mode of sensibility: '*such* a union of these molecules […] which in truth, is just a mode of sensibility'.[33] Forty years later, Fouquet, in his *Discours sur la clinique*, sounds the same theme—Haller 'falsely presented irritability as separate from sensibility, while it is essentially and necessarily related to the former'.[34]

---

[30] Fouquet 1803, 78.

[31] Venel 1765, 909b.

[32] For mysterious reasons Ménuret published mainly under the name Jean-Jacques, although his given name was Jean-Joseph, and his birth date is usually wrongly given as 1733. His Montpellier doctorate in medicine was on biological generation, arguing for epigenesis contra pre-existence (*De Generatione Dissertatione Physiologica*, 1757). Closer inspection of the medical articles in the *Encyclopédie*, notably by Roselyne Rey in her 1987 thesis, published posthumously in 2000, indicated that Ménuret was a major contributor, whose articles display a high degree of intellectual coherence (Jacques Roger and Jacques Proust had called attention to Ménuret earlier). In Rey's view, if we set aside the case of the 'polygraph' Chevalier de Jaucourt, Ménuret's contribution to the medical articles in the *Encyclopédie*, from volume 8 onwards (excluding anatomy, surgery, and the *material medica*) is the largest, most homogeneous set of texts in that work (Rey 2000, 72). His articles span volumes 8–17, and were written between late 1758 and 1761, when he was aged 19–22. Ménuret spent most of his later career as an 'attending physician' at the Montélimar hospital.

[33] Ménuret 1765, 361.

[34] Fouquet 1803, 78–79, N. 5.

Metaphysically, a major step is taken towards the assertion of a 'monistic' ground in which a certain type of matter, organised matter, is alive and senses. Bordeu repeatedly insists that sensibility is neither strictly mechanical nor a property of the soul: it is immanent in living fibres but decentralised and differentiated, since it takes on a form specific to the function of each organ. It is also, he insists along with other *montpelliérains*, 'easier to understand than irritability', and 'can serve quite well as a basis for explaining all vital phenomena, whether in a state of health or of disease'.[35] As much as the vitalists often say that their type of inquiry is neither as reductive as that of the 'mechanists' (the target varies here, sometimes Boerhaave, sometimes the Italian iatromechanists, sometimes even Haller, despite how far removed he is from strict mechanism), nor as supernatural and un-experimental as that of the animists, Bordeu—in this rather different from Ménuret or Fouquet—is willing to tie his originality to Stahlian animism, specifically with regard to sensibility, which he names as the feature common to his, Stahl's, and Van Helmont's models: 'one cannot deny that those who treat each part of the body as an organ or a kind of being or animal with its own movements, action, department, tastes, and particular sensibility drew from the same sources as the Stahlians'.[36]

That sensibility is deliberately being construed as an anti-mechanist concept appears notably with Bordeu's choice of 'model organism', the glands, because their secretory and excretory capacity is precisely the type of function that the mechanist model could not do justice to; they respond to stimuli in ways that mechanism cannot specify, but which are also, of course, independent of soul or will. Bordeu's major work, the *Recherches anatomiques sur la position des glandes et sur leur action* (1752), is devoted to this topic. In this sense harking back to Glisson (who, as shown by Giglioni, was rather more of an active experimenter than Haller gave him credit for), Bordeu wants to stress that the glands have an innate activity and responsiveness to stimuli which can regulate the 'fluid dynamics' of the exchange between the inside and outside of a gland: this property is sensibility. Consistent with the idea that the glands are so many little *lives* (which, however, are independent of the soul), Bordeu also describes this responsiveness as dependent on a kind of *sensation*:

> Secretion can thus be reduced to a kind of sensation, if I may speak thus; the parts that can excite a given sensation will pass through, while the others are rejected; each gland, each orifice will have, so to speak, its personal taste; everything foreign will ordinarily be rejected.[37]

Through this property, fibres, tissues, organs, and organ systems carry out sequences of actions according to what Tobias Cheung has called stimulus-reaction

---

[35] Bordeu 1768/1818, 668.

[36] Bordeu 1768/1818, 671.

[37] Bordeu 1752/1818, Sect. 108, 163. Compare Diderot's, 'Why does each gland have its particular secretion? One cannot really answer otherwise than in terms of irritants, sensibility, animality, taste, the will of the organs' (Diderot 1778/1975–, 387).

schemes.³⁸ For Bordeu, this type of interconnective action is expressed through notions such as 'sympathy' and the 'consensus of the parts', which hark back to the older 'conspiration' (as in Claude Perrault's statement that living bodies differ from 'inanimate bodies' because the former possess 'sympathy and mutual conspiration'; he also speaks of 'commerce' and 'mutual need'³⁹). In this sense, sensibility is also a *network concept*, which easily shows how it can be picked up together with other concepts of the nervous system by a thinker like Diderot—but it is also a strictly *material* concept, without either any intervention of an entity such as the soul, or even of an 'emergentist' conception of hierarchical levels of organisation.

However, there is an ambivalence about the ontological status of sensibility (to borrow an expression from Tobias Cheung's discussion of Bordeu).⁴⁰ That is, generally speaking, sensibility is a property of living matter for Bordeu. And his writing focuses on *medical* entities (rather than questions of basic structure or physiology⁴¹), stressing that the physician is an observer, rather than a quantitative natural philosopher (or experimental physiologist) seeking to discover, say, laws of nervous energy. The physician does not posit the soul, vital principles, or entelechies either. Nevertheless, questions remain. For one, Bordeu, Fouquet, and Barthez in particular speak philosophical language at times (as do Bichat and Bernard in the next generations), but especially, they conceive of sensibility in terms of the property of a substance. Whether or not vitalists are like Stahlians (they often say they are not, but as we saw, Bordeu sometimes equates his sensibility concept with Stahl and Van Helmont), they fall somewhere on this spectrum. Consider this somewhat inflated statement by Charles-Louis Dumas (1765–1813), the Dean of the Montpellier medical faculty in the early nineteenth century, who is defending the Montpellier school in a 'wise', retrospective analysis:

> The various tendencies in medicine stem from philosophers' mistaken applications of the physical sciences or the metaphysical sciences, to the doctrine of living beings. Those who relied excessively on the physical sciences produced the ancient and widespread sect of the materialists. Those who relied on the metaphysical sciences produced the equally ancient sect of the spiritualists. In between these two, there exists a third class of physiologists who do not relate all the phenomena of life to matter or the soul, but to an intermediate principle which possesses properties (*facultés*) different from the one and the other, and which regulates, disposes and orders all acts of vitality, without being impelled by the physical impulses of the material body or the moral affections and intellectual foresight of the thinking principle.⁴²

As a side note, it is interesting that Dumas uses such pure philosophical language to classify trends in medicine. (Claude Bernard also, as I noted, combines philosophical and physiological language, but when he classifies previous doctrines it is

---

³⁸ Cheung 2010, 66–104.
³⁹ Perrault 1680, 201.
⁴⁰ Cheung 2010, 94.
⁴¹ Boury 2008, 528.
⁴² Dumas 1806, Vol. 1, 296, quoted in Rey 2000, 386.

to show how far removed they are from experimental, laboratory science; no such *coupure* here.) But what does Dumas say? That materialists reduce everything to physics, animists (here termed spiritualists) are overly metaphysical, and finally the vitalists, who do things right, do not reduce vital phenomena either to matter or to the soul. What is this vitalist third way (which we need to grasp if we wish to grasp anything distinctive about Enlightenment vitalism)? If Dumas does not say: we vitalists operate heuristically and have understood that, unlike our predecessors, we should bracket off ontological considerations, Barthez does actually say exactly this, in the second edition of his *Nouveaux éléments*, in 1806, in a chapter with the revealing title 'Sceptical considerations on the nature of the vital principle', where he explains that he 'personifies' the vital principle only 'in order to refer to it more easily'; it really has no existence apart from that of the body. And above all, he adds, 'I am wholly indifferent to *Ontology* as the science of entities'.[43]

Bordeu had just such hesitations himself with regard to the ontological status of his 'principle', which he calls sensibility. If we recall that sensibility is often described as a 'self-preserving force' by these authors, that is, a type of reactivity or capacity for responsiveness that ensures our survival (e.g. Fouquet defines it as 'the basis and preservative agent of life'[44] and later, Diderot speaks of sensibility as a 'quality unique to the animal, which warns it of its relations to the surrounding environment'[45]), it is noteworthy that in a key passage of the *Recherches anatomiques*—actually a footnote to what is probably the most famous passage of the book, where he introduces the metaphor of the beeswarm to describe organismic unity—Bordeu asks if the 'ever-vigilant preservative force' that watches over 'all living parts', belongs to '*the essence of a part of matter, or a necessary attribute of its combinations?*'.[46] It is not possible to reconstruct Bordeu's thinking further and provide a definite answer to his question. But we can learn from this that the vitalist doctrine of sensibility poses itself the question, both of the ontological status of this property overall, and of the specific situation that obtains with regard to sensibility as (general) *property of matter* or (more restrictively) *of organisation*.

Vitalist sensibility—Bordeu's and others'—is not a merely mechanical-reaction property, because of its 'network' dimension, its way of explaining and at the same time implying the consensual, sympathetic interaction of the organic parts understood as *little lives*. I have mentioned this idea earlier, but only in passing; suffice it to say, here, that it is a core idea of Montpellier vitalism, consisting in the following:

---

[43] Barthez 1806, Vol. 1, 107, 99, Chap. 3, N. 17, 96 (it can be confusing that the notes added to this edition have their own pagination, also in Arabic numerals: thus the reader can read about the metaphysics of substance on page 96 of the main text and not find 'ontology', but if she turns to later sections where the page numbers restart, these sections appear).

[44] Fouquet 1765, 38b.

[45] Diderot 1778/1975–, 305.

[46] Interestingly—not least for commentators interested in the role of analogy in science—Bordeu here concedes that he must be content here with analogies, 'metaphorical expressions, comparisons'. Bordeu 1752/1818, Sect. 108, 163, Note. Emphasis added.

the organism (living body, animal economy) is not a set of inanimate parts but of organs understood as so many little *lives*. Ménuret speaks of 'the general life formed by the *particular lives* of the organs'[47]; Fouquet says that 'each organ senses or lives in its own way, and the concurrence (*concours*) or sum of these *particular lives* is life in general'.[48] The point was perhaps made best of all in an almost unknown text, a medical thesis on irritability defended at Montpellier in 1776 by a certain Mr 'D.G.' (who further research identifies as Jean Charles Marguerite Guillaume de Grimaud): this applies down to the level of the so-called molecules composing each organ, 'the life of each organ of the animate body is not a simple life, but the real product of as many *particular lives* as there are *living molecules* entering into the composition of the organ'.[49] This is neither mere aggregation of matter, nor mechanical relations between parts defined by shape, size, motion (and position).

However, like irritability, sensibility as discussed here is exclusively material and thus without any 'transcendent' or 'spiritual' dimension.[50] That is, as d'Holbach put it, whether sensibility is 'a quality that can be communicated, like motion, and is acquired through combination', or instead 'a quality inherent to all of matter', in both cases, it cannot belong to 'an unextended being, as the human soul is thought to be'.[51] Further, sensibility has both a reductionist dimension (in this not so far removed from Haller's irritability) and a holistic dimension: the former, because there is a specific analysis of types of tissue, of the structure and function of glands, and so on; the latter, because what is then stressed is the way in which organs interact and produce 'systemic' or 'organisational' properties. The more reductionist vision is apparent when Fouquet, when he underscores the compatibility of Haller's system and the system of sensibility (i.e. his own and Bordeu's), speaks both of the 'consensus of organs' *and* of 'their location' (i.e. spatial, positional information).[52] Or, to take an example from a different, but familiar author, La Mettrie in his 'materialist' rendition of the concept of irritability, also insists that each fibre of animal bodies moves according to an inherent principle, but with a less holistic result than, for instance, what Bordeu (or Diderot) will promote:

> each little fibre, or part of an organised body, is impelled by its own principle, the action of which is not dependent on the nerves, unlike voluntary motions; since these motions occur without the parts involved being in any interaction (*commerce*) with circulation.[53]

What is different in Fouquet's sense of the consensus/conspiration/sympathy of organic parts is that it is a *structural* view. For instance, he also speaks of the

---

[47] Ménuret 1765, 361b.

[48] Fouquet 1765, 42b.

[49] Grimaud 1776, 12 (emphasis in original). I first encountered this text, quoted (only as 'D. G.'; I have added the attribution) in Huneman 2007, 262–276, 390–394 (notes), here, 390, N. 2.

[50] Boury 2008, 529.

[51] d'Holbach 1770/1998–2001, 229–230.

[52] Fouquet 1765, 51a. It is important to remember that articles like these, which came out in the 1765 'batch' of the *Encyclopédie*, are thus 15 years posterior to Bordeu's *Recherches anatomiques*.

[53] This is La Mettrie's comment in *L'Homme-Machine*, after listing ten experiments proving mind-body interaction. La Mettrie 1748, 74, 1960, 181–182.

'economic action of sensibility', with the term 'economic' being reminiscent of the technical term 'animal economy', that is, a system of interdependent relations over and beyond ordinary aggregation of matter, bringing together various 'lives' (active organs) in a manner he describes as 'harmony, symmetry and arrangement'. However, Fouquet—like Bordeu—remains agnostic about whether this harmony, this 'economic action' is the result of *interaction* or just of an additive accumulation of parts ('the concurrence or sum of these particular lives'[54]), closer to La Mettrie's vision.

It is hard to reduce vitalist sensibility to a straightforward claim or set of empirical points, whether we take our bearings for these from the history of medicine, of hybrid discourses 'of the nerves', passions, and spirits, or of course from philosophy. Yet at the same time, the Montpellier vitalists are consistent over time with a set of claims they make with respect to this property, even if they can be more or less Stahlian, more or less Hallerian-compatible, more or less materialism-friendly. There is a general *sensibility monism* here which makes it all the more natural that Diderot found it such an appealing concept—or an appealing medico-theoretical construct to turn into a concept, in order to challenge the Cartesian dualism laid out by the character D'Alembert in the first dialogue of the *Rêve de d'Alembert*.[55]

8.3 Sensibility as a Booster-Property of Matter in Diderot

le vivant et l'animé, au lieu d'être un degré métaphysique des êtres est une propriété physique de la matière[56]

In the very first paragraph of Diderot's 1769 'dialogue' *Le Rêve de d'Alembert*, which was one of his two personal favourites amongst his works (the other being a mathematical essay on probabilities[57]), the character D'Alembert, who is a partisan

---

[54] Fouquet 1765, 42b.

[55] I am not claiming there is some basic, unwavering relation between the 'practice' of physicians and the 'conceptualisation' of a philosopher—here, Diderot. Both because these physicians are very much *médecins-philosophes*, sometimes self-proclaimed, and their writings can bristle with philosophical references (especially Barthez who revised his *Nouveaux elements* with more and more empiricist references, pasting in Bacon and Hume in a desperate hope that his treatise would turn into a perfect piece of empiricism); and of course, because Diderot operates across multiple registers—chiefly, for present purposes, an experimental-naturalistic novel or dialogue, *Le Rêve de d'Alembert*, and a naturalistic proto-work, the *Éléments de physiologie*—which stand in a fertile but ambiguous relation to each other. The well-known fact that Bordeu is also a character in the *Rêve* should illustrate the difficulty of traditional distinctions (without it having to imply that Diderot was the first postmodern, or practitioner of intertextuality).

[56] Diderot and Daubenton 1751, 474a (quoting Buffon, *Histoire générale des animaux*, 'Comparaison des animaux et des végétaux').

[57] Diderot 1955–1961, Vol. 9, 126. Cf. 'Fragments dont on n'a pas pu retrouver la véritable place', in Diderot 1975–, Vol. 17, 223. The *Rêve* was unpublished during Diderot's lifetime (he gave one copy to Catherine the Great as a gift).

of substance dualism, challenges the character Diderot—a materialist, as it happens—to account for the existence of consciousness and thought, and in doing so, introduces the problem of sensibility as a property. Referring to a discussion that seems to have taken place before the text begins, he declares to Diderot, 'this sensibility [...] if it is a general and essential quality of matter, then stones must sense'.[58] That is, if the character Diderot thinks he can successfully defend thinking matter, or a variant of it, by reconfiguring it as *sensing* matter, the character D'Alembert responds: then you will also need to grant that stones can sense. Sensibility is hence present from the first lines of the text, and the word (*sensibilité*) is used a total of 37 times.

How can we define the steps taken from Haller, Bordeu et al. to Diderot? There are two equally trivial ways to proceed, which are roughly symmetrical, and focus respectively on two different works by Diderot, which indeed have a very different status. One is to view Diderot as a kind of proto-Bachelardian poet-metaphysician of the cosmos,[59] as manifest in the *Rêve* with its 'human polyps on Jupiter or Saturn!',[60] and thus present his contribution as a kind of leap into associative freedom beyond the constrained empirical studies of Haller and others. Sometimes this speculative dimension, in which Diderot's scientific imagination can reach conceptual 'places' that science cannot, is described as a kind of science-fiction, or more aptly, as 'a thought experiment on sensibility', in Anne Vila's terms, although she notes that it is a thought experiment which instantly has material effects and conversely, is itself 'materialised'.[61]

The other approach focuses on the *Éléments de physiologie* (an unfinished text on which Diderot worked in the late 1770s), and views Diderot as a commentator on scientific studies of sensibility, who remains at the level of fragments, unable to provide his own scientific theory. Namely, if Haller's physiology contributed the idea of a combinatorial system composed of the structural elements of the organism, which amounted to a system of functional vital properties corresponding to various levels of organic integration,[62] Diderot is, on this view, either a mere commentator on such concepts, or a naturalistically inclined philosopher seeking to accumulate information to support his general vital-materialist views.

A more sympathetic or expansive version of this view, which grants Diderot more originality, is to view his reflections on irritability and sensibility, fibres and organs, bodies and networks as a genuine expansion of vitalist organicism, in the direction of a total 'science of man', understood as an integrated doctrine of the physical and the moral. And it has been observed by commentators at least

---

[58] Diderot 1769/1975–, 90.

[59] As in Alexander 1953 and (in a more sophisticated way) Saint-Amand 1984, where the cosmic dimensions of Diderot's speculations are now justified with quotations on complexity from Michel Serres.

[60] Diderot 1769/1975–, 125.

[61] Vila 1998, 74. For my discussion of this issue see Wolfe 2007, 317–328.

[62] Duchesneau 1999, 197. In the later portions of his article Duchesneau seems to defend Diderot's originality as a contributor to medico-physiological theory.

as far back as Yvon Belaval that the *Éléments*, which Diderot probably intended to publish if he had been able to continue, closely resembles contemporary treatises on 'L'Homme' such as those by Marat or Le Camus.[63] Indeed, there is a careful articulation of Haller, Bordeu, and Barthez in the *Éléments* (along with Whytt and additional figures I shall not discuss); the title itself is, of course, the same as that of the French translation of Haller's 1747 *Primae lineae physiologiae*: *Elémens de physiologie* (first translation by Tarin, 1752, second translation by Bordenave, 1769).

Diderot brings together a mechanistically oriented account of a structural relation between solid parts (from Haller), the more holistic sense of an integrated network of sensibility/sympathy (from Bordeu and Barthez), and various other theories of organic matter concerning what we might call 'vital *minima*', that is, the minimal constituents of organic life which are themselves 'alive' and possessed of animate properties.[64] And he collapses any residual dualist distinction between irritability and sensibility (which after all, in Haller and in Whytt, although in completely different ways, had served to preserve a concept of soul): 'In general, in the animal and in each of its parts—life, sensibility, irritation'.[65] Differently from Whytt or Bordeu and his colleagues who had to insist on a quasi-metaphysical primacy of sensibility, Diderot just renders them identical:

> This force of irritability is different from any other known force; it is life, sensibility; specific to the soft fibre; weaker, then extinct in the tightened fibre; greater in the fibre attached to the body than to the fibre separated from it. This force is not dependent on gravity, attraction or elasticity.[66]

The life of the 'whole animal', is the composite of the life of each organic component, interacting in a relation of 'sympathy', which sometimes is not dependent on any centre, any 'controller' at all: 'there are sensing and living organs, coupling, sympathising and concurring towards the same goal, without the participation of the whole animal'.[67] This raises the question of the *unity* of the organism (in the *Rêve*, the unity of the self, which Mlle de Lespinasse worries about—to which the character Bordeu replies precisely with a doctrine of *organismic* unity, that is, you are yourself because of the individuality of your body or *organisation*). After all, if an organism is a sum of many lives, whether this is an additive sum or one that involves qualitative shifts, where is the limit? This is another one of the difficult questions which neither Diderot nor Bordeu—both of whom pose it—resolve to anyone's satisfaction, including their own. One recalls that Bordeu introduced the image of the beeswarm as a *metaphor* of organic unity, and Diderot, although he expands on it and adds other metaphors including the spiderweb and the harpsichord (for the vibrating 'strings' of the nervous system), does not present it as anything other than that. Now, my purpose

---

[63] Belaval 2003, 257.
[64] Wolfe 2010, 38–65.
[65] Diderot 1778/1975–, 449.
[66] Diderot 1778/1975–, 308.
[67] Diderot 1778/1975–, 501.

here is not to reconstruct a possible 'materialist theory of the self', in Diderot and others,[68] but rather to enquire into the extent to which a concept like sensibility functions as a 'booster' for the materialist—a *functional booster*, at the level of physiology and medicine, and an *ontological booster*, with respect to levels of organisation, emergence, and reduction.

Yet we must not lose sight of the fact that this appropriation of the concept of sensibility is a key part of Diderot's attempt to articulate organic unity, as something different from the unity of machines, or that of the universe as a whole. And, crucially for the specifically *biomedical* context I have sketched, this attempt is not generically metaphysical or inspired by classic texts in the history of philosophy, but is particularly close to medical texts such as Bordeu's; as Henry Martyn Lloyd suggests in Chap. 9, 'for the discourse of sensibility, the master discourse was medicine'. Even if the idea of matter as possessing animate features can be viewed as something Leibnizian (as a 'materialisation of the monad', as it is sometimes described, or a kind of panpsychism, which La Mettrie had already recognised as a danger: 'the Leibnizians, with their *Monads*, put forth an unintelligible hypothesis. They have spiritualised matter rather than materializing the soul'[69]), or as harking back to Renaissance matter theory as in Campanella,[70] it has a very particularly medical, *embodied* flavour here. Consider Diderot's approach to the unity of animals or organisms he calls 'continuity' as opposed to the merely spatial 'contiguity' that exists between heaps of matter: 'Without sensibility and the law of continuity in animal substance (*contexture*), without these two qualities, the animal cannot be one'.[71] Biology and medicine or metaphysics were hard to separate with respect to sensibility as late as the mid-nineteenth century, as noted by Littré: 'sensibility or the function of the nerves [...] is a final terrain in which theology and metaphysics still compete with biology'.[72]

That sensibility is a medical concept with an expansive conceptual potential can also be seen in another way: Diderot (and partly La Mettrie before him, for whom 'irritability' is the general monistic term rather than 'sensibility') sees that a concept such as sensibility allows him to integrate conceptually the reactivity and representational capacity of mind (the nervous system, the brain as a 'book which reads

---

[68] I attempt an initial presentation of the problem in Wolfe 2011a.

[69] La Mettrie 1748, 2; 1960, 149.

[70] Diderot provides some indication as to the Leibnizian provenance of his idea of sensibility as a universal property of matter in his *Encyclopédie* entry 'Leibnitzianisme', where he associates Aristotelian entelechies, monads, and 'sensibility [as] a general property of matter' (Diderot 1765b, 371a). As in other cases, his source is Johann Jakob Brucker's 1744 *Historia critica philosophiæ*. Belaval notes that the publication of Jean-Baptiste Robinet's Leibnizian *Philosophie de la nature* in 1765—the year of the letter to Duclos—may have led Diderot to the idea of consideration of animate parcels of matter (Belaval 2003, 334, N. 3). For more on the Leibnizian background of sensibility, see Nakagawa 1999, 199–217. Jean Varloot sees the notion of a universal sensibility in matter as going back all the way to Campanella! (in Diderot 1962, Vol. 3, ci, N. 3).

[71] Diderot 1778/1975–, 307.

[72] Littré 1846, 229.

itself', as Diderot puts it[73]) *while maintaining a thoroughgoing naturalism*—there are no properties which are not properties of natural beings subject to causal processes as specified in the natural sciences (whatever these may be: thus the naturalism of a Hobbes or a d'Holbach, who seem to be intuitively *physicalists*, is very much a reduction to the physical properties of matter, while the naturalism of a Gassendi, a Diderot, or, a few decades later, an Erasmus Darwin is a reduction to matter conceived as the bearer of vital, animate properties, typically attributed to minimal components of matter named 'semences', 'seminarerum', or 'molecules'[74]). This naturalism has been interpreted in various ways by Diderot commentators in recent decades: as 'monism', or 'holism', or again 'emergentism'. These contemporary terminological decisions do not modify the fundamental intuition that (i) matter is 'one', a unified whole (both at the level accessible to our measuring instruments and at a metaphysical level: Nature makes no leaps), (ii) properties such as sensibility, consciousness, memory, desire, instinct are 'just there'—no room for external world scepticism, 'no pleasure that is felt is chimerical'[75]—and as such belong to the material whole as stated in (i).

Diderot is less willing to commit to a definitive position regarding (iii) whether these properties are *universal properties of matter*, as he often says (we might also say 'basic properties', thinking of Ménuret's insistence that movement and sensibility reduce to 'one primitive notion'[76]), or properties only of *organised wholes*: 'Sensibility, a general property of matter or a product of organisation'.[77] The first view certainly fulfils the requirements of a materialist metaphysics, and is pleasingly immanentist, except that it is also a potentially 'panpsychist' view in which tiny parcels of matter are themselves said to think, feel, remember, and react (recall La Mettrie's warning about 'spiritualising matter'); the second view offers the advantage of a hierarchical arrangement in which there are levels of organisation—today we might say 'levels of complexity'—which are interrelated within a general material whole.

Here we leave specifically Bordevian or vitalist territory in Diderot and return to metaphysics. In the earlier *Pensées sur l'interprétation de la nature* (1753), Diderot had reflected on the quasi-aporia of the relation between living matter and dead matter, and put forth a series of 'queries' (somewhat reminiscent of the Queries which followed Newton's *Opticks*) which tended to challenge the

---

[73] 'The soft substance of the brain [is] a mass of sensitive and living wax, which can take on all sorts of shapes, losing none of those it received, and ceaselessly receiving new ones which it retains. There is the book. But where is the reader? The reader is the book itself. For it is a sensing, living, speaking book, which communicates by means of sounds and gestures the order of its sensations'. (Diderot 1778/1975–, 470.)

[74] On vital *minima* in a materialist context see Wolfe 2010.

[75] Diderot, Letter III to Falconet, in *Le pour et le contre (correspondance avec Falconet)*, in Diderot 1975–, Vol. 15, 9.

[76] Ménuret 1765, 361b.

[77] Diderot 1769/1975–, 105.

distinction between these two states. Whether we view this as an empirical or a metaphysical issue in Diderot, he definitely insists that the distinction is false inasmuch as what is alive is constantly in a process of fermentation and corruption, and what is dead is conversely in a process of being *assimilated* into life, like the marble of the statue, ground into earth, growing into plants, and eaten by animals, and so on—a process for which he or d'Holbach coined a term, 'animalisation'. As Diderot says in his marginal commentary on Franz Hemsterhuis's *Lettre sur l'homme* (1773–1774):

> When I was born, I could only sense along a length of about eighteen inches at the most. How was I able, with time, to feel along a length of five feet and some inches? I ate. I digested. I animalised. By a process of assimilation, I turned *corps bruts* from inert to active sensibility.[78]

So animalisation is a process which ensures that matter is sensible, since it is constantly moving from inert to active; and the new distinction between *inert* sensibility and *active* sensibility can help resolve some of the above difficulties.[79] But isn't this just another version of dead matter versus living matter? Or (to point to a different problem), monism seems to indicate that Diderot should opt for *one kind of matter*, not two, and then claim that this matter senses. But—as he notices in his critique of Maupertuis' panpsychism—it seems to be a mistake (although of what sort is not clear: empirical? metaphysical?) to endow the element—the 'molecule'—with the properties of the whole—*l'organisation*, or here, to endow matter with the properties of organised wholes.[80]

Again, what is the status of sensibility? Diderot's dilemma, or at least his ontological decision, returns here: 'Sensibility, a general property of matter or a product of organisation'.[81] He addresses this in a variety of texts—'speculative' ones such as the *Rêve*, 'experimental' ones such as the *Éléments* (however much the distinction between speculative and experimental may be shopworn and of limited use here), letters to Sophie Volland (October 1759) and better-known, to Duclos (October 1765), commentaries and critiques on other thinkers such as Hemsterhuis and Helvétius. Before trying to achieve some resolution on the issue by way of conclusion, let me try and map out the situation in Diderot.

First, there is no clear-cut distinction between different texts which represent different positions on the issue, as some have suggested. Granted, the *Rêve* is more speculative than the *Éléments*, but even in the latter, he asks, 'Why not consider sensibility, life and motion as so many properties of matter, since these qualities are to be found in every portion, every particle of flesh?'[82] Yet, second, it is clear that different viewpoints are adopted, not some kind of perpetual

---

[78] Diderot, *Observations sur Hemsterhuis*, in Diderot 1975–, Vol. 24, 304.
[79] See the brief but useful discussion in Duflo 2006, 347–352.
[80] Wolfe 2010, 57, 65.
[81] Diderot 1769/1975–, 105.
[82] Diderot 1778/1975–, 333.

polyphony. Thus in the *Réfutation d'Helvétius*, 4 years after the *Rêve*, Diderot calls the general sensibility of matter a mere 'supposition', which is not sufficient for 'good philosophy', and admits that 'the necessary connexion in this shift, escapes me'.[83] That is, how can inert matter become active matter? This is why epigenetic processes such as embryo growth in the egg are so metaphysically 'pregnant', so to speak, for Diderot: because they provide evidence that out of exclusively material layers something like life (a.k.a. sensibility) emerges. Hence sciences such as the nascent biology of the eighteenth century but also chemistry and medicine are of great importance, if not in filling out the blanks in this 'passage' so that all necessary causal links are made explicit, at least in articulating it as a material process.

If Diderot's 1759 and 1765 letters treat us to some real phantasmagorias, with the idea of matter possessing sensation for all eternity[84] so that the molecules of lovers buried side by side will join together after the deaths of their individual organisms, and less romantically, the description of the animal as a 'laboratory' in which sensibility shifts from inert to active,[85] in the last texts, including the *Eléments*, the problem of whether sensibility is a universal property of matter indeed becomes strictly experimental, with considerations of flayed vipers, the trunks of eels, and sectioned grass snakes:

> I am inclined to believe that sensibility is nothing other than the motion of animal substance, its corollary; if I introduce torpor, i.e., the end of movement at a given point, sensibility also ceases. [...]
> The sensibility of matter is the specific life of the organs. The proof of this is obvious in the viper that has been skinned and beheaded; in the section of the eel and other fish, in the grass snake divided into parts, in the various separate, palpitating parts of the body, in the contraction of the heart when it is pricked.[86]

And he explicitly uses the language of 'demonstration': 'Someday it will be demonstrated that sensibility or touch is a sense common to all beings. Some phenomena already indicate this'.[87] Sensibility as the life 'proper to organs', as a sense which is 'common to all beings': we are back at a vitalist vision of sensibility as the life of a system of organs.

---

[83] Diderot 1875/1994, 297–298.

[84] 'Feeling and life are eternal. What lives has always lived and always will. The only difference I know between death and life is that at present, you live as one mass, and that once dissolved, scattered into molecules, twenty years from now you will live in detail.' (Diderot, letter to Sophie Volland, 17 October 1759, in Diderot 1955–1961, Vol. 2, 283–284.)

[85] Diderot, letter to Duclos, 10 October 1765, in Diderot 1955–1961, Vol. 5, 141.

[86] Diderot 1778/1975–, 305–306. Jacques Chouillet suggests that Diderot wrote the *Principes philosophiques sur la matière et le mouvement* (1770) to resolve empirically this problem of the 'passage' from the inert to the active, even though there is no discussion of sensibility or life in the text; but there is another, even more fundamental act of 'monistic collapse', of the difference between inertia and motion (Chouillet 1984, 54).

[87] Diderot 1778/1975–, 308.

8.4 Conclusion

Diderot adopts a vitalist solution to the series of metaphysical aporias concerning sensibility and matter. He did not opt for the straightforward solution that sensibility results from organisation, instead stating, even when discussing particulars such as grass snakes (much as La Mettrie had combined the metaphysics of irritability with the case of lizards in one passage), that sensibility is a property of matter. As Timo Kaitaro has put it, if sensibility results from organisation, one then has the problem of explaining this organisation, whereas if matter possesses some vital properties, an elementary form of sensibility, this could be used in explaining its tendency to form organised wholes.[88] And, as we saw with respect to Fouquet, Bordeu, and La Mettrie's versions of the organism as composed of little lives, there are also different degrees in their articulations of organisational 'wholeness'. The aporias of living and dead matter, inert and active sensibility, and generally, sensibility as ontologically irreducible or as a result of certain types of organisation, may or may not be fully resolved, even if it is hopefully clear that the subtle vitalist reflections did not just arrive at Diderot's 'science-fiction' or phantasmagorias as a terminus. But we have seen that the property of sensibility acts as a conceptual booster—the materialist's privileged route of access to 'what lies higher', as seen with the dialogue between Diderot and D'Alembert—and one which is of specifically medical origin.

In the end, for Diderot, rocks do not sense except in the rather 'God's-eye', Spinozist sense that in the long run, they too will be 'animalised'. Organisms sense; sensibility is the definitory property of organic matter. Thought cannot result from the mere spatial proximity of molecules, the contiguity of matter; it 'results from sensibility', which is inert in *corps bruts* like rocks, and active in living bodies, by being assimilated with 'living animal substance'.[89] In addition to the rather technical considerations we have encountered concerning how organisms hang together, it is important to remember that if D'Alembert grants Diderot's claim that matter can sense—that sensibility is a universal or general property of matter—he will have granted everything, for Diderot, in this extending an empiricist insight which nowhere appears as radically as in his version, has collapsed all cognitive functions into modes of sensation: 'The only thing that is *innate* is the faculty of sensing and thinking; all the rest is acquired', or as d'Holbach has it in the *Système de la nature* (a work on which Diderot was an active collaborator), 'What is it to think, enjoy or suffer, if not to sense?'[90]

I have not tried, as is often done, to reconstruct a problem and its solutions, such as, 'how did these thinkers move from a mechanistic model to one recognising the complexity of sensibility as a feature, either of the self-regulation of organisms and/or of the nervous system?' Rather, I have suggested a topography of the problem of sensibility as property of matter or as vital force in the mid-eighteenth

---

[88] Kaitaro 2001, 113.
[89] Diderot, letter to Duclos, 10 October 1765, in Diderot 1955–1961, Vol. 5, 141.
[90] Diderot 1765a, 754a; d'Holbach 1770/1998–2001, 322.

century, in the tangle of disciplines and discourses devoted to the nature of living, biological entities—not an exhaustive cartography of all possible positions or theories, but an attempt to understand the 'triangulation' of three views: a mechanist, or 'enhanced mechanist', view in which one can work upwards, step by step from the basic property of irritability to the higher-level property of sensibility (Haller); a vitalist view, in which sensibility is fundamental, matching up with a conception of the organism as the sum of parts conceived as little *lives* (Bordeu et al.); and, more eclectic, a materialist view which seeks to combine the mechanistic, componential rigour and explanatory power of the Hallerian approach, with the monistic and metaphysically explosive potential of the vitalist approach (Diderot). As we have seen, the relation between the medical-vitalist approach to sensibility and Diderot's appropriation and transformation of that approach, is not one that lets itself be labelled easily, although his conceptual innovation in developing what Anne Vila calls the 'superproperty' of irritability and sensibility taken as a whole,[91] is undeniable. In the 'laboratory' of the animal which forms the metaphysical horizon of the embodied materialist, 'to sense is to live'.[92]

Acknowledgments Thanks to Alexandre Métraux for his critical remarks.

References

Alexander, Ian. 1953. Philosophy of organism and philosophy of consciousness in Diderot's speculative thought. In *Studies in romance philology and French literature presented to John Orr*, 1–21. Manchester: Manchester University Press.
Barthez, Paul-Joseph. 1806. *Nouveaux éléments de la science de l'homme*, 2nd ed, 2 vols. Paris: Goujon & Brunot.
Belaval, Yvon. 2003. *Études sur Diderot*. Paris: PUF.
Bordeu, Théophile de. 1752/1818. Recherches anatomiques sur la position des glandes et sur leur action. In *Oeuvres complètes, précédées d'une Notice sur sa vie et ses ouvrages par Monsieur le Chevalier de Richerand*, 2 vols., vol. 1, 45–208. Paris: Caille et Ravier.
Bordeu, Théophile de. 1768/1818. Recherches sur l'histoire de la medicine. In *Oeuvres complètes, précédées d'une Notice sur sa vie et ses ouvrages par Monsieur le Chevalier de Richerand*, 2 vols., vol. 2, 548–734. Paris: Caille et Ravier.
Boury, Dominique. 2008. Irritability and sensibility: Key concepts in assessing the medical doctrines of Haller and Bordeu. *Science in Context* 21(4): 521–535.
Campanella, Tommaso. 1672. *De natura substantiae energetica*, 187, cited in Giglioni 2008, 479.
Cheung, Tobias. 2010. *Omnis Fibra Ex Fibra*: Fibre architectures in Bonnet's and Diderot's models of organic order. *Early Science and Medicine* 15: 66–104.
Chouillet, Jacques. 1984. *Diderot poète de l'énergie*. Paris: PUF.
Condillac, Etienne Bonnot de. 1754/1984. *Traité des sensations et Traité des animaux*. Paris: Fayard.

---

[91] Vila 1998, 15.
[92] Diderot 1778/1975–, 447.

d'Holbach, Paul-Henri-Thiry, Baron. 1770/1998–2001. *Système de la Nature ou des lois du monde physique et du monde moral*. In *Oeuvres philosophiques*, 3 vols., vol. 1, ed. J.-P. Jackson. Paris: Editions Alive.

Diderot, Denis. 1765a. Inné. In *Encyclopédie ou Dictionnaire raisonné des arts et des métiers*, 35 vols., vol. 8, ed. Denis Diderot and Jean Le Rond D'Alembert, 754. Paris: Briasson, David, Le Breton & Durand.

Diderot, Denis. 1765b. Leibnitzianisme. In *Encyclopédie ou Dictionnaire raisonné des arts et des métiers*, 35 vols., vol. 9, ed. Denis Diderot and Jean Le Rond D'Alembert, 369–379. Paris: Briasson, David, Le Breton & Durand.

Diderot, Denis. 1769/1975–. *Le Rêve de d'Alembert*. In *Oeuvres complètes*, 25 vols., vol. 17, ed. Herbert Dieckmann, Jacques Proust, Jean Varloot et al., 25–209. Paris: Hermann.

Diderot, Denis. 1778/1975–. Eléments de physiologie. In *Oeuvres complètes*, 25 vols., vol. 17, ed. Herbert Dieckmann, Jacques Proust, Jean Varloot et al., 261–574. Paris: Hermann.

Diderot, Denis. 1875/1994. Réfutation suivie de l'ouvrage d'Helvétius intitulé *l'Homme*. In *Philosophie*, vol. 1, ed. Laurent Versini. Oeuvres. Paris: R. Laffont, collection "Bouquins".

Diderot, Denis. 1955–1961. *Correspondance*, 9 vols., ed. Georges Roth. Paris: Éditions de Minuit.

Diderot, Denis. 1962. *Oeuvres choisies*, 3 vols., ed. Jean Varloot. Paris: Editions sociales.

Diderot, Denis. 1975–. *Oeuvres complètes*, 25 vols., eds. Herbert Dieckmann, Jacques Proust, Jean Varloot et al. Paris: Hermann.

Diderot, Denis, and Louis Jean-Marie Daubenton. 1751. Animal. In *Encyclopédie ou Dictionnaire raisonné des arts et des métiers*, 35 vols., vol. 1, ed. Denis Diderot and Jean Le Rond D'Alembert, 468–474. Paris: Briasson, David, Le Breton & Durand.

Duchesneau, François. 1982. *La physiologie des lumières. Empirisme, modèles, théories*. The Hague: Nijhoff.

Duchesneau, François. 1999. Diderot et la physiologie de la sensibilité. *Dix-huitième siècle* 31: 195–216.

Duflo, Colas. 2006. Sensibilité. In *L'Encyclopédie du Rêve de d'Alembert de Diderot*, ed. Jean-Claude Bourdin, Colas Duflo, et al., 347–352. Paris: Éditions du CNRS.

Dumas, Charles-Louis. 1806. *Principes de physiologie, ou introduction à la science expérimentale, philosophique et médicale de l'homme vivant*, 2 vols. Paris: Crapelet.

Fouquet, Henri. 1765. Sensibilité, Sentiment (Médecine). In *Encyclopédie ou Dictionnaire raisonné des arts et des métiers*, 35 vols., vol. 15, ed. Denis Diderot and Jean Le Rond D'Alembert, 38–52. Paris: Briasson, David, Le Breton & Durand.

Fouquet, Henri. 1803 [an XI]. *Discours sur la clinique*. Montpellier: Izar & Ricard.

French, Roger K. 1969. *Robert Whytt, the soul and medicine*. London: Wellcome Institute.

Garrett, Brian. 2003. Vitalism and teleology in the natural philosophy of Nehemiah Grew. *British Journal for the History of Science* 36: 63–81.

Giglioni, Guido. 2008. What ever happened to Francis Glisson? Albrecht Haller and the fate of eighteenth-century irritability. *Science in Context* 21(4): 465–493.

Grimaud, Jean-Charles-Marguerite-Guillaume de ['D.G.']. 1776. *Essai sur l'irritabilité*. Avignon: Bonnet frères.

Haigh, Elizabeth L. 1984. *Xavier Bichat and the medical theory of the eighteenth century*. London: Wellcome Institute for the History of Medicine.

Huneman, Philippe. 2007. 'Animal economy': Anthropology and the rise of psychiatry from the *Encyclopédie* to the Alienists. In *The anthropology of the enlightenment*, ed. Larry Wolff and Marco Cipolloni, 262–276, 390–394 (notes). Stanford: Stanford University Press.

Kaitaro, Timo. 2001. 'Man is an admirable machine'—A dangerous idea? *La lettre de la Maison française d'Oxford*, special issue on *Mécanisme et vitalisme* 14: 105–121.

La Mettrie, Julien Offray de. 1748. *L'Homme-Machine*. Leiden: E. Luzac.

La Mettrie, Julien Offray de. 1960. *L'Homme-Machine*, ed. A. Vartanian. Princeton: Princeton University Press.

Littré, Émile. 1846. De la physiologie (review essay on Johannes Müller's *Manuel de Physiologie*). *Revue des deux mondes* 14: 200–237.

Ménuret de Chambaud, Jean-Joseph. 1765. Œconomie Animale (Médecine). In *Encyclopédie ou Dictionnaire raisonné des arts et des métiers*, 35 vols., vol. 9, ed. Denis Diderot and Jean Le Rond D'Alembert, 360–366. Paris: Briasson, David, Le Breton & Durand.

Nakagawa, Hideo. 1999. Genèse d'une idée diderotienne: la sensibilité comme propriété générale de la matière. In *Être matérialiste à l'âge des lumières, Hommage offert à Roland Desné*, ed. Beatrice Fink and Gerhardt Stenger, 199–217. Paris: PUF.

Perrault, Claude. 1680. Expériences pour l'Éclaircissement de la circulation de la sève des plantes. In *Essais de physique ou recueil de plusieurs traitez touchant les choses naturelles*, 4 vols., vol. 1, 195–255. Paris: Jean-Baptiste Coignard.

Reill, Peter Hanns. 2005. *Vitalizing nature in the enlightenment*. Berkeley: University of California Press.

Rey, Roselyne. 2000. *Naissance et développement du vitalisme en France de la deuxième moitié du 18e siècle à la fin du Premier Empire*. Oxford: Voltaire Foundation.

Roe, Shirley A. 1984. Anatomia animata: The Newtonian physiology of Albrecht von Haller. In *Transformation and tradition in the sciences: Essays in honor of I. Bernard Cohen*, ed. Everett Mendelsohn, 274–300. Cambridge: Cambridge University Press.

Saint-Amand, Pierre. 1984. *Diderot et le labyrinthe de la relation*. Paris: Vrin.

Sonntag, Otto (ed.). 1983. *The correspondence between Albrecht von Haller and Charles Bonnet*. Bern: Huber.

Steinke, Hubert. 2005. *Irritating experiments: Haller's concept and the European controversy on irritability and sensibility, 1750–1790*. Amsterdam/New York: Rodopi.

Venel, Gabriel-François. 1765. Irritabilité. In *Encyclopédie ou Dictionnaire raisonné des arts et des métiers*, 35 vols., vol. 8, ed. Denis Diderot and Jean Le Rond D'Alembert, 909. Paris: Briasson, David, Le Breton & Durand.

Vila, Anne C. 1998. *Enlightenment and pathology. Sensibility in the literature and medicine of eighteenth-century France*. Baltimore: Johns Hopkins University Press.

von Haller, Albrecht. 1747/1765. *Primae lineae physiologiae*, 3rd revised & expanded ed. Göttingen: Ap. Vid. Ae. Vandenhoeck.

von Haller, Albrecht. 1755/1936. *A dissertation on the sensible and irritable parts of animals*. London: J. Nourse; reprint, Baltimore: The Johns Hopkins Press.

von Haller, Albrecht. 1756–1760. *Mémoires sur la nature sensible et irritable des parties du corps animal*, 4 vols. Lausanne: Bousquet.

von Haller, Albrecht. 1777. Sensibilité. In *Supplément à l'Encyclopédie, ou dictionnaire raisonné des sciences, des arts et des métiers, par une société de gens de lettres*, 4 vols., vol. 4, 776a–779b. Amsterdam: M.-M. Rey.

von Haller, Albrecht. 1779. *First lines of physiology*. Trans. from the correct Latin edition printed under the inspection of William Cullen. Edinburgh: Printed [by Macfarquhar and Elliot] for Charles Elliot.

Wellman, Kathleen. 2003. Materialism and vitalism. In *The Oxford companion to the history of modern science*, ed. J.L. Heilbron. Oxford: Oxford University Press. *Oxford Reference Online*. http://www.oxfordreference.com. Accessed 15 Sept 2011.

Whytt, Robert. 1768. *The works of Robert Whytt*. Edinburgh: Beckett and Balfour.

Wilkes, John (ed.). 1810. *Encyclopaedia Londinensis, or, Universal dictionary of arts, sciences, and literature*, vol. 1. London: Adlard.

Wolfe, Charles T. 2007. Le rêve matérialiste, ou 'Faire par la pensée ce que la matière fait parfois'. *Philosophiques* 34(2): 317–328.

Wolfe, Charles T. 2011a. Éléments pour une théorie matérialiste du soi. In *La Circulation entre les savoirs au siècle des Lumières. Hommages à Francine Markovits*, ed. François Pépin, 123–149. Paris: Hermann.

Wolfe, Charles T. 2011b. From substantival to functional vitalism and beyond, or from Stahlian animas to Canguilhemian attitudes. *Eidos* 14: 212–235.

Wolfe, Charles T. 2010. Endowed molecules and emergent organization: The Maupertuis-Diderot debate. *Early Science and Medicine* 15: 38–65; reprinted in T. Cheung, ed. 2010. *Transitions and borders between animals, humans and machines, 1600–1800*. Leiden: Brill.

Chapter 9
Sensibilité, Embodied Epistemology, and the French Enlightenment

Henry Martyn Lloyd

Abstract This chapter reconstructs the theory of knowledge as it operated in the French Enlightenment. It does so initially by questioning the extent to which epistemology was divided between 'British empiricism' and 'Continental rationalism', and by showing that in the discourse of sensibility, if the theory of knowledge was 'first philosophy', then it was so in terms largely set by Enlightenment vitalism. Building on these initial points, the chapter opens with an examination of the interaction between medical vitalism and sensibility, where the latter is understood as both a passive and an active power of the living body. Here, I begin to tease out, not what is continuous between Locke and the French Enlightenment, but what was *added to* Locke's thought by the period. In the second section, I examine the implications of this understanding of the body of sensibility for what has been called the period's 'philosophical particularism' and for its practice of science. Here, the body of sensibility was constructed as always particular. The ability of the theory of sensibility to constitute a unifying ground within a discourse which produced a proliferation of particularity is the focus of this section. The chapter moves from considering the body of sensibility as the object of knowledge to considering it as the subject that knew.

It is a still predominant, if now much criticised, view of early-modern philosophy that it reached its zenith with Kant. From the towering achievement of the first *Critique's* (1781) 'synthesis' of 'British empiricism' and 'Continental rationalism', a meta-narrative has been imposed retrospectively on all that came before it. As Knud Haakonssen has argued, this 'epistemological paradigm' is grounded in the idea that 'the theory of knowledge is at the core of all sound philosophy': that

---

H.M. Lloyd (✉)
Centre for the History of European Discourses and School of History, Philosophy,
Religion and Classics, The University of Queensland, 4072 St Lucia, QLD, Australia
e-mail: m.lloyd@uq.edu.au

epistemology constitutes 'the true *prima philosophia*'.[1] The broad structure of this meta-narrative is very familiar:

> The epistemological approach divided post-Renaissance philosophy into two major schools or directions, namely, rationalism, and empiricism. The former had commonly been seen as characteristic of the European continent, though one of the defining features of eighteenth-century philosophy, on this view, was that France gradually switched from Cartesian rationalism to Lockean empiricism, embodied by Condillac. Germany, however, was supposed to maintain a continuous development of rational system-building through Leibniz, Wolff, and their followers and opponents. In contrast, the English-speaking world was seen to pursue the empiricist view in ever-finer detail from Bacon to Hobbes through Locke, Berkeley, and Hume.[2]

Note Condillac's place in this story; it is broadly accepted that Condillac 'adopted Locke's empiricism as the basis of his own philosophy'.[3] Note too, that Condillac is taken here to be representative of French epistemology as a whole. Haakonssen continues by noting that

> the epistemological paradigm for early-modern philosophy has been an immensely powerful vehicle for scholarship and for the self-understanding of the discipline of philosophy. Nevertheless, the paradigm is arguably at considerable variance with the philosophical self-understanding common in that period, and this [...] suggests that it is part of the philosophical historian's task to question it.[4]

Such questioning can take two forms: it can involve, first, the extent to which epistemology was in fact divided between something like 'British empiricism' and 'Continental rationalism', and second, the extent to which the theory of knowledge was in the period 'first philosophy'.

In moving to reconstruct the theory of knowledge as it existed in the French Enlightenment, this chapter will participate in both these modes of questioning. It will do so initially by noting two general points. First, it is only a very superficial understanding of the terms 'rationalism' and 'empiricism' that fixes them into a mutually exclusive binary. Nuanced understandings recognise that the two terms may each identify different and complementary features of a single theory of knowledge. Accordingly, research into the details of views actually held invariably reveals a mélange of archetypically 'rationalist' and 'empiricist' views. This is particularly the case for France, which the epistemological paradigm itself sees as moving from Cartesian 'rationalism' to Lockean 'empiricism' over the period in question. The difficulty of separating 'rationalism' from 'empiricism' in Enlightenment discourses is exacerbated by the fact that the two terms were actually part of *nineteenth-century* reconstructions of the period and not part of its self-understanding. Further, the two terms do not map cleanly onto the terms which the period *did* use—for example, 'experimental' and 'speculative' natural philosophy—and even if they did, an exclusory binary is still not evident. Epistemology over the period

---

[1] Haakonssen 2006, 7.
[2] Haakonssen 2006, 7.
[3] Knight 1968, 8. See also Yolton 1991, 4, 72–74, 210.
[4] Haakonssen 2006, 13.

was in a 'state of flux' and this was reflected in what Peter Anstey has described as a 'vagueness or indeterminacy' in even those categories which were part of the period's self-understanding.[5]

Second, even without forcing the theory of knowledge to take on the structure of an exclusive binary, by imposing on the period an idea of what constitutes 'proper' epistemology, the 'epistemological paradigm' obscures the importance for the history of French philosophy of the discourse of sensibility. It obscures, that is, the discourse's *actual* theory of knowledge. For, while I do think the theory of knowledge was a very important part of the foundation of much of the period's philosophy, it was not the case, as the epistemological paradigm implies, that either the French Enlightenment broadly construed or the discourse of sensibility more narrowly understood knowledge in *merely* Lockean terms. It is particularly important to note here the movement away from the mechanist/corpuscularian matter theory which underpinned Locke's epistemology. Mechanistic matter theory was in decline in the mid- to late-eighteenth century and there were at least two main responses to this decline: neo-mechanism (as found, for example, in D'Alembert, Condorcet, Lagrange, and Laplace) and Enlightenment vitalism.[6] As I argue in the introduction to this volume, vitalist medicine was one of the central features of the discourse of sensibility. This was particularly the case in France in the second half of the eighteenth century. As this chapter will show, the change to a vitalist theory of matter had important effects within an ostensibly Lockean framework.[7] If in the discourse of sensibility as it was manifest in France in the second half of the eighteenth century, the theory of knowledge was 'first philosophy', then it was so in terms largely set by Enlightenment vitalism and by the *médecins philosophes*. At least here, for the discourse of sensibility, the master discourse was medicine[8]; in the terms of the *Encyclopédie*, 'this science is more important than any other'.[9] It is hard to overestimate the influence of medicine for the period and for the *philosophes*: for the *Encyclopédie*, to quote Anne Vila, 'the enlightenment truly *was* a medical matter', and particularly for the 1765 volumes, medicine specifically meant vitalist medicine.[10]

---

[5] Anstey 2005, esp. 220, 238.

[6] Reill 2005, 5–7, 33–70. See also Gaukroger 2010, 387–420.

[7] John Yolton, in the most comprehensive text on Locke in eighteenth-century France, did not recognise the significance of vitalist medicine for the theory of knowledge in the period. For example, in his brief entry on Le Camus, he quickly noted the continuities between him and Locke without commenting on the significant differences in matter theory which underpinned the 'medical men's' interest in physiology. (Yolton 1991, 15, 68–69.) While Yolton did not mention vitalism, he did devote a chapter to the place of the physiological/medical in the period's move towards materialism. (Yolton 1991, 86–109.) His focus in this text was the metaphysics of mind and body. His text then had difficulty bringing sensibility into focus (as sensibility did not necessarily imply materialism and was in the period invoked by both dualists and materialists.) For a broad history of the change in matter theory see Gaukroger 2010.

[8] See Morris 1990.

[9] Anonymous 1765a, 315. My thanks to Kim Hajek for providing the translations.

[10] Vila 1998, 80. See also Moravia 1978.

Building on these initial points, this chapter opens with an examination of the interaction between medical vitalism and sensibility understood as both a passive and an active power of the living body. I will here begin to tease out, not what is continuous between Locke and the French Enlightenment, but what was *added to* Locke's thought by the period—notably, a theory of active matter—and the effects of this addition. In the second section, I examine the implications of this understanding of the body of sensibility for what has been called the period's 'philosophical particularism' and for its practice of science. Here, the body of sensibility was constructed as always particular, with the degree and nature of sensibility differing not only between various parts of the body, but also between particular individuals. The ability of the theory of sensibility to constitute a unifying ground within a discourse which produced a proliferation of particularity is the focus of this section. The chapter moves from considering the body of sensibility as the object of knowledge to considering it as the subject that knew; I explore this, firstly, in terms of the Enlightenment's much-touted 'rational' subject, and secondly in terms of the 'empirical' subject.

9.1 Medical Vitalism and Embodied Epistemology

The major reference to sensibility in the *Encyclopédie* was the major (almost 17,000-word) article 'Sensibilité, Sentiment (Médecine)'.[11] The article was written by Henri Fouquet, a minor figure associated with the faculty of medicine at Montpellier and an acquaintance of Diderot and D'Alembert, whom he had met at the *Collège Royal* and the *Jardin du Roi* in Paris.[12] Although Fouquet's article was categorised 'Médecine', it encompassed themes which were unselfconsciously metaphysical in nature. The article and its placement in the *Encyclopédie* were important because they demonstrated the manner in which the discourse of sensibility underpinned and unified topics which prima facie may be thought to have been quite disparate. There was an astonishing breadth in the speculative concept elucidated in the rather ecstatic tones of the article's opening: for Fouquet, sensibility or sentiment was 'the faculty of feeling, the principle of sensitivity, or the very feeling of the parts, the basis and conserving agent of life, animality par excellence, the most beautiful, the most singular phenomenon of nature'.[13] For Fouquet, sensibility was *in*, or a property of, a living body; it was that which preserves life. Diderot, elsewhere in the *Encyclopédie*, defined sensibility as that which opposes death[14]; sensibility simply *was* life.

Fouquet's article was consistent with a self-standing text on a similar theme, Antoine Le Camus's *La Médicine de l'esprit* (1769).[15] Le Camus, named *docteur*

---

[11] Fouquet 1765.

[12] Dulieu 1952.

[13] '*la faculté de sentir, le principe sensitif, ou le sentiment même des parties, la base & l'agent conservateur de la vie, l'animalité par excellence, le plus beau, le plus singulier phénomène de la nature*'. Fouquet 1765, 38.

[14] '*La mort n'est que la cessation de la sensibilité*'. Diderot 1755, 782.

[15] Le Camus 1769. This second edition differs substantially from the first edition published in 1753.

régent of the conservative Paris faculty of medicine in 1745 and appointed to the chair of surgery in 1766, showed in this text the extent to which 'Montpellier' vitalism had, by the 1760s, penetrated French medical thought.[16] The text clearly illustrated what it was to be a *médecin philosophe*: the first of its three books, *La Logique des médecins*, surveyed the metaphysical foundations of medical theory, focusing particularly on the understanding/*entendement* and the will/*volonté*, and on causes in general, including the physical causes, which influenced the mind.[17] Commencing with a brief introduction on understanding ('the general faculty of knowing (*connaître*)'[18]), a discussion which rapidly deferred to Locke ('chief of the *Philosophes*'[19]), and which noted that understanding consists of sense and reflection (which importantly ought not to be understood as independent of corporeal motions), the text rapidly arrived at the first substantive chapter, 'De la Sensibilité & des Sensations'.

> Before knowing, it is necessary to feel; before feeling, it is necessary to be sensitive. It is thus necessary to speak of sensibility before examining the sensations, which are the origin of our knowledge. A difficult subject, but worthy of research by any *Philosophe*. While one need not go out of oneself to grasp it, one must have pondered on the whole of nature to treat it pertinently.[20]

Or as he explained elsewhere, 'sensibility and the sensations [are] the simplest properties of our bodies, which contribute the most to the operations of understanding (*l'entendement*), and [are] necessarily linked to them'.[21] Sensibility was the 'force of all our knowledge (*connaissances*), just as it was the source of all our passions'.[22] 'Force' here was understood as the tonic force (*force tonique*) and was differentiated from the elastic and muscular forces. Le Camus also used the term 'vital force' (*force vitale*) to describe it.[23]

> This force is a continual tendency to shortening, sometimes even an actual shortening. Its action is inseparable from life, only lasts while life remains and is the first principle of sensibility.[24]

---

[16] Vila 1998, 81. See also Rey 2000, 252–255.

[17] Le Camus 1769, Vol. 1, 10.

[18] Le Camus 1769, Vol. 1, 15.

[19] Le Camus 1769, Vol. 1, 15.

[20] '*Avant de connoître il faut sentir; avant de sentir il faut être sensible. Il est donc nécessaire de parler de la sensibilité avant d'examiner les sensations que sont le principe de nos connoissances. Matiere difficile, mais digne des recherches de tout Philosophe. Si l'on ne doit pas sortir de soi-même pour la saisir, il faut avoir médité sur toute la nature pour en traiter pertinemment.*' Le Camus 1769, Vol. 1, 19.

[21] Le Camus 1769, Vol. 2, 83.

[22] Le Camus 1769, Vol. 2, 84.

[23] It is worth noting that Le Camus uses the term *force vitale* to describe *sensibilité* where Fouquet tends to use the term *flamme vitale*. Le Camus 1769, Vol. 1, 24; Fouquet 1765, 39, 41. For further discussions on the metaphysics of eighteenth-century vitalism, see Charles Wolfe's chapter in this volume (Chap. 8). See also Kaitaro 2008; Wolfe 2012. For Senebier's use of the term *forces vitales*, see Marx 1974, 213.

[24] Le Camus 1769, Vol. 1, 21–22.

The force particularly pertained to animals and to the sensible parts of the body, but not, for example, to bones. The causes which elicited the response of tonic fibres were either exterior or interior impressions, the latter being the passions. The tonic force emanated from the nerves and gave sensibility to the whole body. And so, for Le Camus too, 'sensibility only lasts as long as life, and life only lasts as long as tonic action persists'.[25]

There were two main aspects to the concept of sensibility—it was both passive and active—and Fouquet, in particular, made this explicit. In the first instance, sensibility was essentially passive: 'a power reduced [from potential] to an action, [...] it essentially consists of a purely animal intelligence, which discerns the use or the harm in physical objects'.[26] In this sense, as the power to discern external objects, for Fouquet, sensibility was the ontological basis of sensations ('*sentiment, sensatio, sensus*'). For Le Camus too, sensibility was the precondition of the ability to sense passively: 'sensibility is the aptitude to receive impressions from objects'.[27]

> We say that a sentiment is an impression excited in the soul by sensations, and that sensations are affections of the body caused by a change which occurred on the occasion of a movement produced by the presence of objects, or [sensations are] equivalent to that which is excited by the presence of objects.[28]

By defining sensibility in this way, the narrow philosophical idea of sensation was brought within its scope and Fouquet's article incorporated the much shorter articles 'Sens (Métaphysique)' and 'Sensations (Métaphysique)', articles which constituted relatively orthodox presentations of Locke's theory of sensation.[29] In terms of the ongoing tradition of Lockean empiricism, the ostensible orthodoxy of these two articles may well give the impression that, at least at this point, the French Enlightenment *was* more or less faithfully Lockean. And to a point it was. But there was something else going on, too. Note that neither sense nor sensation had ontological primacy here; sensibility did, and the ability to receive passively sensations of external objects was only one of sensibility's two major aspects.

Sensibility's second major aspect was its activity: 'action or *mobility*, is only the mute expression of this same *sentiment*, that is, the impulsion which carries us towards these objects, or away from them'.[30] The examples that Fouquet gave here were of 'lower' animals. The point was important; animals, even simple ones, 'dilate themselves, open out, so to speak, draw themselves up, become aroused (*eriguntur*), at the approach of objects that they recognise as useful to them, or which pleasantly flatter their *sensibility*'.[31] Sensibility here was a responsive power

---

[25] Le Camus 1769, Vol. 1, 34.
[26] Fouquet 1765, 38. Contrast Fouquet's opinion with Rousseau's, as discussed by Cook in Chap. 5.
[27] Le Camus 1769, Vol. 1, 19.
[28] Le Camus 1769, Vol. 1, 35.
[29] Anonymous 1765b, c.
[30] Fouquet 1765, 38. Emphasis in original.
[31] Fouquet 1765, 38. Emphasis in original.

and as such, it extended well beyond the five passive senses.[32] In fact, for the discourse of sensibility, the 'five' senses were not necessarily privileged: they were only some of a very great variety of centres and types of sensibility (and arguably several of the less important). Specifically, for Fouquet, the heart and diaphragm, or 'epigastric region', was one of the 'primary centres' of sensibility and had a sensibility particularly associated with the passions and the moral sense. In sensing pleasure, the sensitive soul/*âme sensitive* agreeably moved, widened, swelled, and the feeling of pleasure spread. In sorrow, sadness, or terror, the soul temporarily withdrew to the core of the body.[33] There was continuity here with articles such as 'Sens moral, (*Moral*.)', articles which, without Fouquet's article showing the concept which subsumed them, may today be read as having nothing in common with the metaphysical articles on sense or sensation.[34] Notably, moral sense was defined in terms continuous with those of the body of sensibility:

> Moral sense [...] the name given by the *savant* Hutcheson to that faculty of our soul, which, in certain cases, discerns promptly good and moral evil by a kind of sensation and by taste, independently of reasoning and of reflection.
> It is this that other *moralistes* call the *moral instinct*, a sentiment, type of penchant or natural inclination which brings us to approve certain things as good or laudable, and to condemn others as bad and reprovable, independently of all reflection.[35]

Le Camus made this explicit too:

> [Sensibility] is that which gives rise to tenderness for relatives, pity for the destitute, piety towards the Creator, friendship for one's fellows, love for [someone of] different sex, humanity towards one's neighbour, gratitude towards benefactors, resentment against affronts, respect for virtue.[36]

The manner in which moral sense was construed as a 'movement of the heart', or 'interior sensation', which operated independently of reflection, is significant.[37] This chapter will return to this theme.

For Fouquet, then, sensibility had active and passive aspects. Very importantly, however, he noted that the difference between these aspects was the work of the imagination alone: it was 'in this double relationship of actions so closely linked together, that only the imagination can follow them or distinguish them, that *sensibility* should be considered, and its phenomena assessed'.[38] This union of what otherwise might be thought of as separate categories—passive reception of sensation and active response—was fundamental to the discourse of sensibility as it was influenced by vitalist medicine. But rather than addressing the breadth of this idea across the period, and in order to begin to mark explicitly the differences

---

[32] See Singy 2006.
[33] Fouquet 1765, 40, 42. See also Le Camus 1769, Vol. 1, 45.
[34] Jaucourt 1765, 28.
[35] Jaucourt 1765, 28. Emphasis in original.
[36] Le Camus 1769, Vol. 2, 85.
[37] Jaucourt 1765, 28.
[38] Fouquet 1765, 38. See also Boury 2008, 523.

between the thought of the period and Locke's, I want to show its importance in relation to the emblem of French epistemology, Condillac, specifically to the *Traité des sensations* (1754).

In brief, Condillac sought to show, in his elaborate philosophical fiction, that a statue-man, possessing just the faculty of sensation and without the (Lockean) faculty of rationality, could develop a mind.[39] Condillac's statue was 'organised on the inside like us', though initially deprived of the five external senses; but the statue was alive, and to be alive, here, was to have sensibility.[40] Condillac was building on vitalist presuppositions at least in this minimal sense.[41] And so in keeping with sensibility's two aspects, when given senses, the statue was not just given the power to sense external objects, it also had the power to remember or retain those sensations. And it had the power to respond. This active power was the grounds for the arousing of the statue's passions. For Condillac, sensations were inherently either pleasant or painful. This produced desires and these desires led to the development of rational thought-processes including abstract ideas.[42]

As far as the *philosophes* were concerned, Condillac was relatively conservative; not so Helvétius, who realised and developed the radical potential of the *Traité*. Partly in order to introduce themes which will be central to the second half of this chapter, and partly in order to make manifest the relationship between this chapter and broader themes in the volume, it is worth noting that the discourse of sensibility contained within it anxiety over the implications of sensibility and contestation as to its meaning. Though several examples could be invoked here, I will use the case of Diderot and his 1758 *Réflexions sur le livre de l'esprit*.[43] Helvétius's *De l'Esprit* was an extravagant exultation of the body of sensibility, the usefulness of passions, and the perfectibility of the mind. The text made visible the implications of Condillac's ostensibly conservative *Traité*. Building on the sensationist idea that sensations were either pleasant or painful, Helvétius argued that individuals and groups made judgements based on the agreeableness of impressions and the utility or disutility of ideas. Personal interest dictated the judgement of individuals. Collective interest dictated the judgement of groups.[44] Concomitantly, Helvétius was enormously optimistic about the perfectibility of the mind: genius, he argued, was an effect of education rather than a gift of nature.[45] Where the pessimism of Samuel-Auguste Tissot was a medical-hygienic counterweight to Helvétius's

---

[39] Condillac 1754/1970, 10.

[40] Condillac 1754/1970, 39.

[41] Condillac's presupposition is shared by Hume, though perhaps in a more minimal sense: in Hume, without the metaphysical apparatus of the vital force, vitalism is simply a craving for mental exercise which 'puts the Humean mind in motion'. (Cunningham 2007, 61.) See also Rey 1995, 279–280; Rey 2000, 405–407.

[42] Condillac 1754/1970, 45–47, 70–71, 74. This is consistent with Le Camus 1769, Vol. 1, 398–411.

[43] Diderot 1758/1875–1877.

[44] Helvétius 1758, 46–48, 54.

[45] Helvétius 1758, 251, 473–474.

optimism, Diderot's response was predicated on traditional philosophical concerns. Diderot began by showing that if, as Helvétius took it to be, 'sensibility is a general property of matter', then what followed was the collapse of the faculties: 'to perceive, to reason, [and] to judge, is to feel'.[46] Though, as Charles Wolfe shows in Chap. 8, Diderot elsewhere *celebrated* the possibility of sensibility as a general and essential property of matter, here, he worried about the implications of this idea: 'considering the mind (*esprit*) relative to error and to truth, M. Helvétius convinces himself that there is no false mind (*esprit*)'[47]; truth was established by the extent to which something was useful or interesting. For Diderot, this failed to recognise the equivalence between uninteresting and useless geometrical scribbling and the grandeur of the Newtonian laws of celestial bodies; in both cases 'the sagacity is the same, but the interest is another matter, as is the public esteem'.[48] Continuing this theme for moral considerations, Diderot worried that, for Helvétius, 'there is neither justice, nor absolute injustice', and that he did not understand that there was 'an eternal basis to what is just and unjust'.[49] Finally, for Diderot, Helvétius ignored the fact of natural variations which were not subject to willed change, that 'a slight alteration in the brain reduces the genius to a state of imbecility', and that *l'homme de génie* and *l'homme ordinaire* may develop from the same cause.[50] Broadly then, the point is that even in the period's progressive philosophy, there was anxiety caused by the implications of the reduction of the faculty of rationality to sensibility.

Students of Locke will find much in Condillac that they recognise (if not necessarily in Helvétius's radicalisation of him); it is not accidental that the period is understood to be, and perhaps is dismissed as, Lockean. I want to draw attention here not to what was continuous between Locke and the French Enlightenment, but to what was *added* to Locke by the period and the effects of this addition. Notwithstanding his famous hypothesis of thinking matter, Locke gave an account of matter which was both mechanistic and corpuscular.[51] In contrast, in the French Enlightenment, living matter (sometimes *all* matter) was understood in terms of vitalism and sensibility. The point was that ascribing active properties to matter (i.e. the *âme sensitive*), even where the *âme raisonnable* was not itself thought of in materialist terms (as neither Fouquet nor Le Camus did), had the effect of muddying the ontological waters. This produced effects for the period's theory of knowledge which have been little discussed. Partly this is because they lead to a change in emphasis, if not in philosophical substance; Locke did, even if relatively briefly, consider sensation in the form of pleasure and pain, and this was linked by him to the response of the passions.[52] And it was within the conceptual space of Locke's

---

[46] Diderot 1758/1875–1877, 272. See also Diderot 1758/1875–1877, 267.
[47] Diderot 1758/1875–1877, 268.
[48] Diderot 1758/1875–1877, 269.
[49] Diderot 1758/1875–1877, 272.
[50] Diderot 1758/1875–1877, 270–272.
[51] Yolton 1991, 38; Tipton 1996, 78–81.
[52] Locke 1690/1849, Book 2, Chap. 20.

'thinking matter' that were situated the broad series of vitalist responses to mechanism which were a major part of the French Enlightenment and which have been documented elsewhere, including in chapters in this volume.[53] Yet this change in emphasis was significant for a number of reasons. First, it oversaw a breakdown in what may retrospectively be thought of as rigid genre divisions, allowing, for example, the novel to be understood as a properly philosophical/scientific genre. As I will argue below, within the discourse of sensibility, especially as it was manifest in France, the medical/scientific genius was understood in ways which were highly proximate to, perhaps even indistinguishable from, the artistic/moral genius. Second, it allowed the medicalised body to develop a very particular epistemological importance. And third, the rhetoric of the period elevated the particular; the discourse of sensibility worked to universalise the particular and to celebrate genius (whether philosopher, natural scientist, etc.) in its particularity.

9.2 Philosophical Particularism

The implications of this medico-philosophical anthropology were considerable, particularly for what Jessica Riskin has called the period's 'philosophical particularism' and for the practice of science.[54]

As it was printed in the *Encyclopédie*, Fouquet's article is 14 pages long. To this point, I have concentrated only on the first three pages, and on considerations which Fouquet himself designated as being 'purely metaphysical' and 'speculative'.[55] He then clearly marked a move to discussions of particular observations of phenomena, the 'particular effects of sensibility', observations which constituted the major part of his article.[56] That is, after discussing in metaphysical terms the power which unifies them, the article focused mainly on the sensitive body's specific, distinctive, diverse, and unusual phenomena. Sensibility was the unifying concept behind the diversity of the five external senses and accounted for the differences between them: the eyes are sensitive to light, the ears to sound; the eyes see, they do not hear.[57] And sensibility was the unifying grounds of much greater diversity than this. I have already mentioned the sensibility of the heart and its implications for moral sense. Digestion, too, depended on the particular sensibility of the stomach. Importantly, this particularity did not just differentiate the sensibility of the stomach from other regions of the body, but differentiated one stomach from another, with reference to the taste/*goût* of particular stomachs. Fouquet ascribed this variation to differences in secretions. The emphasis here is on 'remarkable variations'. Further, differences

---

[53] See, too, Yolton 1991.
[54] Riskin 2002, 145.
[55] Fouquet 1765, 40.
[56] Fouquet 1765, 40.
[57] On the various tastes of the organs and the three main centres of the body's sensibility see Cheung 2008.

in temperament were regarded as the result of differently modified organs and of the habits associated with their operation.[58] This trope was absolutely fundamental to the *Encyclopédie* and to the thought of the period more broadly. The following is from Le Roy's article 'Homme (Morale)':

> The faculty of feeling probably belongs to the soul, but it only exercises its functions through the intervention of the material organs, which together make up our body. From this arises a natural difference between *men*. The tissue of fibres not being the same in all, some must have certain organs which are more sensitive (*sensible*), and consequently, must receive from objects which affect [those organs], an impression whose force is unknown to others. Our judgements and our choices are but the result of a comparison between the different impressions we receive. They are thus as little alike from one *man* to another as those very impressions. These variations must give to each *man* a kind of particular aptitude which distinguishes him from others by his inclinations, just as he is [distinguished] from the outside by his facial features. Hence, we can conclude that our judgement of others' conduct is often unjust, and that the advice we give them is even more often useless. My reason is inaccessible (*étrangère*) to that of a *man* who doesn't feel in the same way as I do, and if I take him for a madman, he has every right to regard me as an imbecile.[59]

The notion that the natural diversity of humans was founded on embodied difference was a significant theme, too, for Le Camus; 'the human mind (*esprit*) is a real chameleon, tak[ing] on all the colours of the objects which surround it'.[60]

> There is nothing isolated (*désuni*) in nature. Everything is linked to everything; and man, this being whose pride would separate him from the others, is so strongly united to the air, to fire, to the earth, that he ceases to be if he is separated from these elements which keep him alive, which contribute to his health, and which modifying his body in different ways, [and which] must necessarily modify his mind (*esprit*) in different ways.
>
> Everything which produces, surrounds, or maintains our body, can thus bring about notable changes in our souls.[61]

Throughout the second book of the treatise, Le Camus expanded at great length on differences in *la génération* (i.e. inherited traits), sex, climate, seasons, education, temperament (an analysis which continued the traditional classification in accord with the four humours), *le régime de vivre* (which included diet, exercise, sleep, and so forth), age, and general health. This list of influencing factors seems relatively stable across texts. For its part, Fouquet's article extrapolated on differences in 'sensibility at different ages [and] in different sexes', in the 'quality of the air and the impressions of some other external bodies', on the 'influence of the stars', and on 'sensibility in relation to climate'.[62] Similarly, for Le Camus:

> The same senses in different individuals are organised in diverse ways, which makes them susceptible to pleasure or to pain on receiving the same impressions. The music which pleases some, displeases others; some colour agreeable to one, is detested by another;

---

[58] Fouquet 1765, 41–45.

[59] Le Roy 1765, 275. Emphasis in original. This idea is repeated in the article Sensations (Métaphysique), Anonymous 1765d, 25.

[60] Le Camus 1769, Vol. 1, 383.

[61] Le Camus 1769, Vol. 1, 291.

[62] Fouquet 1765, 46–49.

> somebody eagerly seeks out a given smell, whereas somebody else avoids it with horror. Dishes are more or less delicious, more or less bad, for different palates. Age, which changes all constitutions, at the same time, changes the way of feeling of the same organs in the same individuals. From this it follows that tastes change, and that we no longer have the same affections. [...] It is for all these reasons that we can say that every organised being has its own way of feeling.[63]

Here we can begin to see the way the category of sensibility diversified and that it did so in decidedly un-Lockean ways. This diversification is paralleled in Fouquet's discussion of the particular effects of music and the experience of beauty, both of which derive from 'a disposition of the organs, a matter of taste in the feeling soul (*l'âme sensitive*) which is affected in this or that manner'.[64] This understanding of what it was to sense was carried across to other articles in the *Encyclopédie*. Specifically, the fact that our sentiment or sensibility was agreeably affected by the beautiful and the harmonious was a feature of the articles 'Sens (métaphysique)' *and* 'Plaisir (morale)'.[65] The examples given were indicative of the types of judgement which were at stake: they focused on experiences of colour or music which pleased one but displeased another. They did not focus on perceptions which differed in terms of the hue or pitch of the colour or notes. Such a distinction was a feature, too, of Jean Senebier's *L'Art d'observer* (1775),[66] a text which 'arguably represents the formalised sum of the eighteenth century art of observation',[67] and which will be discussed in more detail below. I want to mark the link to theories of moral sense, as this feature was key to the discourse of sensibility; sense, and the diversity of sensations, were understood in terms of what we may today think of as the observer's aesthetic/moral response to a phenomenon, rather than as objective or instrumental. Again, I will develop this point below.

Within the discourse of sensibility, the focus on embodied particularity was summarised well by Senebier:

> The same causes therefore produce in each People and in each man the same effects, provided that we note the modifications that each of these causes can receive from the particular state of each People or of each man, and the circumstances in which they can exist.[68]

The sensible body, understood as the object which was produced by the discourse, was constructed as particular and diverse. Significantly—and this is the point at which the period's theory of knowledge becomes central to the considerations of this chapter—the object of the discourse was also the subject who knew; the

---

[63] Le Camus 1769, Vol. 1, 46–47.

[64] Fouquet 1765, 45.

[65] Anonymous 1765b, c, 689, 691.

[66] Senebier 1775, Vol. 1, 111–112. This is also a feature of the article Sensations (Métaphysique), Anonymous 1765d, 24–25.

[67] Singy 2006, 54. See also Marx 1974, 205; Legée 1991.

[68] Senebier 1775, Vol. 2, 251–252. Senebier's list of particulars which must be noted when observing a society include climate, government, religion, the state of the sciences, and the state of the women. Senebier 1775, Vol. 2, 222–239.)

period's knowledge-seeker was also understood in terms of the discourse of sensibility and hence in particular terms.

Without implying that there was an exclusive disjunction in operation—rather, as I stress in the introduction to this chapter, one of the fundamental ideas here is that, in the epistemology of the French Enlightenment, the two coexist—the remainder of this chapter will illustrate the implications of this conception of the knowledge-seeker for what can be understood as the period's 'rational' and 'empirical' subjects.

In the first instance, the focus on embodied particularity had significant implications for the Enlightenment's much touted 'rational' subject. Le Camus made it very clear that he was not a materialist, that he was 'not unaware that the soul is a contingent, rational, spiritual, and immortal substance'.[69] Specifically, he was an occasionalist.[70] Belief in a 'rational and immortal soul' was maintained by Fouquet.[71] So too by Senebier, who 'regretted the atheist orientation taken by French philosophy, and thought that the only true *philosophy* must serve as a basis for Christianity, not undermine its foundations'.[72] There is no reason to think this (relative) orthodoxy was disingenuous; the rational soul did play a role here, even if it was mostly a formal role. I have above noted the anxiety which Helvétius's *De l'Esprit* caused Diderot, and I noted that his response was predicated on traditional philosophical concerns and showed a genuine nostalgia for a traditionally construed faculty of rationality. The immortal soul provided stability, transportability, and, literally for Le Camus, 'homogeneity'.[73] However, this emphasis on the stabilising role of the rational soul was in stark contrast to themes of the particularity and malleability of the body, which were the main concern of the discourse of sensibility. In thematic terms, the idea of the rational/immaterial soul in fact played a very small role: 'if we consider that God must have created souls as essentially the same, as His goodness encourages us to believe, souls must only be modified differently through their union with the body'.[74] Consequently, Le Camus wrote, that notwithstanding his knowledge of the immaterial rational soul, 'I also know that due to truly mechanical causes, the soul is helped or constrained in its operations, that often due to causes of the same nature, it is diverted in its functions independent of its will'.[75] His emphasis in the text was then not on the universality of reason, but on the particularity of the body and its effects, including its particular effects on reasoning. Of five people with one (different) sense each, he wrote: 'they would have two sentiments in common, pleasure and pain, but they would still reason differently on the nature of these

---

[69] Le Camus 1769, Vol. 1, xv.
[70] Le Camus 1769, Vol. 1, 93; Yolton 1991, 69.
[71] Fouquet 1765, 39.
[72] Marx 1974, 210.
[73] Le Camus 1769, Vol. 2, 403.
[74] Le Camus 1769, Vol. 1, 7–8.
[75] Le Camus 1769, Vol. 1, xv.

general and universal modes'.⁷⁶ This fact was emphasised in the most overt statement of his occasionalism:

> God alone is the efficient cause of our ideas, because He is the only being capable of producing movement by himself and of acting on minds (*esprits*) and on bodies; but God only excites ideas in our souls according to the dispositions of our bodies: the dispositions of our bodies are thus the occasional causes of our ideas.⁷⁷

Similarly, Fouquet allowed a place for the rational soul by separating the manner by which our organs attained knowledge from our intellectual judgements:

> From this [the habits of organs] can also arise this animal movement [which is] always founded on the habits of our sensibility, renewed by its instinct in the presence of an object dear to us, and which a change in the [external] features masks from our intellectual habits; such is the situation of a loving mother in the presence of a son she no longer recognises, and towards whom, nevertheless, her sensitive soul seems to want to fly.⁷⁸

But again, the focus was on differentiated sensibility rather than universalised rationality, and Fouquet proceeded to separate humans from animals not in terms of their rationality, but rather in terms of their superior sensibility: Compared to animals and compared to one other, it was superior sensibility which led to superior understanding. If humans possessed a higher degree of intelligence compared to animals, it was not because of their rational soul, but because of the fact that they 'possess[ed] sensibility to the highest degree'.⁷⁹

That is, the raison d'être of both Le Camus's and Fouquet's texts was premised on the problem of embodied particularity occasioned by the variability of sensibility, and not on the reliability and stability of the rational soul.⁸⁰ Thus, where Descartes's *Regulae ad direction emingenii* (*Rules for the Direction of the Mind*) foregrounded the question of the 'mind's eye' (Rule 5) over the *actual* eye, and Locke's *An Essay Concerning Human Understanding* focused almost exclusively on what we may call mind-based solutions to the problem of knowledge (Locke adopted from Descartes the 'way of ideas'⁸¹), for Le Camus, the solution to the problem of knowledge was embodied and hygienic.

> We have, or so we think, sufficiently proven the power of climate, of education, both moral and physical, of lifestyle, of temperaments, of seasons, etc., on the mind. In elaborating on the way of acting of all these causes, we have, at the same time, seen how much they contribute to the diversity of [the quality of] genius, of characters, virtues, vices, passions and morals. It is on these principles that we establish the power of Medicine on the soul, and the power of the Physician to regulate the penchants and the animal functions of men. [... W]e

---

[76] Le Camus 1769, Vol. 2, 119. See also Senebier 1775, 98–99.
[77] Le Camus 1769, Vol. 1, 93.
[78] Fouquet 1765, 45.
[79] Fouquet 1765, 46.
[80] See also Senebier 1775, Vol. 2, 211.
[81] Yolton 1990.

deduce from these [principles], the physical and mechanical means of rectifying the defects of the mind, of increasing its capabilities, and of preserving its good qualities.[82]

After having reflected attentively on the physical causes which, modifying bodies in different ways, also alter the dispositions of minds [in various ways], I was convinced that by employing these different causes or by imitating their power with [our] art, we would succeed in correcting defects (*vices*) of understanding and the will by purely mechanical means.[83]

While maintaining a formal ontological separation between mind and body, sensibility was nonetheless understood to be the functional ground of both the passions and the mind. That this was the case is even clearer for Condillac; although he, too, denies being a materialist, one of the dominant features of Condillac's thought was his attempt to eliminate the faculty of rationality and maintain only one faculty, that of sense.[84]

Notwithstanding authors' continued reliance on the immaterial and rational soul, there was an unambiguous move here toward a materialist theory of mind. In Chap. 8, Charles Wolfe has written at length about Diderot's materialism[85]; we may also name figures such as Helvétius and d'Holbach. This tendency was noted by conservative contemporary commentators, including critics of Le Camus and of Condillac.[86] As I noted above, it was the cause of Diderot's anxieties vis-à-vis Helvétius. The ever-increasing emphasis on the importance to the mind of physiological considerations and of embodied particularity—and therefore of philosophical particularity—which I have been tracing here, led contemporary critics to react against what was seen as the three interconnected ills of the progressive philosophy of the period, 'materialism, fatalism, and Pyrrhonism'.[87] In tracing the period's philosophical particularity, this chapter has traced some of the links between what was held to be an increasing tendency towards materialism and a concomitant Pyrrhonian attack on the possibility of knowledge.

Second, the period's focus on embodied particularity had significant implications for the Enlightenment's 'empirical' subject. Specifically, I want to draw attention to the implications for the medical/scientific observer. Vitalist medicine operated not only in terms of the particular sensibility of the patient, but in terms of the particular sensibility of the physician or observer. Now, as we have seen,

---

[82] '*Nous avons, à ce que nous pensons, suffisamment prouvé la puissance des climats, de l'éducation tant morale que physique, du régime de vivre, des tempéraments, des saisons, &c, sur l'esprit. En développant la manière d'agir de toutes ces causes, nous avons vû en mêmes-tems combien elles contribuoient à la diversité des génies, des caracteres, des vertus, des vices, des passions & des mœurs. C'est sur ces principes que nous établissons le pouvoir de la Médecine sur les ames, & le pouvoir du Médecin pour regler les penchans & les fonctions animales des hommes. [... N]ous en déduirons les moyens physique & méchaniques de rectifier les défauts de l'esprit, d'en augmenter la mesure & d'en conserver les bonnes qualités*'. Le Camus 1769, Vol. 2, 54–55.

[83] Le Camus 1769, Vol. 1, vi–vii.

[84] Condillac 1754/1970, 10.

[85] See also Wolfe and Terada 2008, 565–568; Wolfe 2012.

[86] Note Yolton's particular reference in this context to texts by Roche and Boullier, in Yolton 1991.

[87] See Yolton 1991, 73, 111.

sensibility was not held to be a uniform property, neither within nor between observers. The consequences for the observer were evident in Ménuret de Chambaud's article 'Observateur, (Gram. Physiq. Méd.)'.[88] Ménuret's article did contain praise of the empirical observer, much of which was ostensibly Lockean. But it needs to be noted that observation of nature was not conceived here as operating within a mechanistic ontology, and the epistemological stability and transportability provided by the faculty of rationality were little in evidence. Rather, observers were characterised in terms of their highly cultivated sensibility.

The article sets the observer in firm opposition to the experimental natural philosopher, who 'never sees nature as it is in reality, he claims, through his work, to make it more appreciable, to remove the mask which hides it from our eyes, [but] he often disfigures it and makes it unrecognisable'.[89] It also criticises the mere natural historian, marking as incorrect the title 'observer' given to 'the ignorant *empirique*, the humdrum practitioner, the preoccupied systematiser, the compiler of observations, the describer of illness'.[90] It is significant that the article was hostile to those who founded their practice on 'rules [which are] always general, and never [on] particularities'.[91] Unencumbered by speculative principles, by abstract learning, the observer was open to the always *particular* signs of nature. In this sense,

> [The observer] follows nature step-by-step, reveals the most secret mysteries, everything strikes him, everything informs him, all results are just the same to him because he does not expect any of them, with the same eye, he discovers the order which reigns over the entire universe and the irregularity to be found there; for him, nature is a great book which he has only to open and consult; but to read this immense book requires genius and penetration, it requires lots of insight; to do experiments requires only adroitness: all the great *physiciens* were observers.[92]

But this revelation was not readily transferable: observers had to learn for themselves how to read the signs of nature, something which could not be done through systematic or transferred learning, but individually, by the bedside of the ill. Consequently, the observer was characterised in highly particular terms and as a heroic individual—a *'génie observateur'*—endowed with a highly developed

---

[88] Ménuret 1765, 310–313; Vila 1998, 52–65. Ménuret features heavily in Rey's study as *'représentant exemplaire des vitalistes qui ont collaboré a l'*Encyclopédie'. (Rey 2000, 60.)

[89] *'ne voit jamais la nature telle qu'elle est en effet, il prétend par son travail la rendre plus sensible, ôter le masque qui la cache à nos yeux, il la défigure souvent & la rend méconnoissable'*. Ménuret 1765, 310.

[90] Ménuret 1765, 313.

[91] Ménuret 1765, 313.

[92] *'suit pas-à-pas la nature, dévoile les plus secrets mysteres, tout le frappe, tout l'instruit, tous les résultats lui sont égaux parce qu'il n'en attend point, il découvre du même oeil l'ordre qui regne dans tout l'univers, & l'irrégularité qui s'y trouve; la nature est pour lui un grand livre qu'il n'a qu'à ouvrir & à consulter; mais pour lire dans cet immense livre, il faut du génie & de la pénétration, il faut beaucoup de lumieres; pour faire des expériences il ne faut que de l'adresse: tous les grands physiciens ont été observateurs'*. Ménuret 1765, 310. See, too, Senebier 1775, Vol. 1, 5–6. Senebier also speaks in detail about *adresse*. (Senebier 1775, Vol. 1, 131–135.)

'*finesse dans le sentiment*' (keenness of feeling).[93] 'The aptitude to succeed (*talent*) as an *observer* is more difficult than one would think'[94]; 'The designation of *observer* is an honourable title in Medicine, which is, or rather should be, the lot of the physician'.[95]

These themes introduced by Ménuret were developed by Jean Senebier in his *L'Art d'observer*. The text was, in the main, a detailed manual of natural philosophy devoted to questions of induction, analogical reasoning and the analytic method, general laws of nature, and the nature and use of hypotheses. As did Le Camus and Ménuret,[96] Senebier represented the problem of the development of the arts and sciences in terms of self-cultivation. For Le Camus, this was hygienic cultivation; for Senebier, though he does hold that 'it is firstly necessary that the senses of the Observer be well-constituted, that is, that each be in a fit state to yield its full effect',[97] the emphasis was more heavily on the moral. In large part, the treatise focused on the qualities/*adresse* which the observer had to cultivate: patience and especially attention, which rendered the observer *pénétrant* (of penetrating in mind), *exact* (rigorous or precise), and which assured the quality of observations.[98] Much of the text proceeded in terms which were to become increasingly prominent in the emergence of nineteenth-century science: Senebier recognised that while 'it is almost impossible to observe the same thing twice in the same way', this did not preclude a 'theory of the certainty of observations'.[99] Part of Senebier's response to the problem of particularity was to rely on a corporate notion of observations, and he invoked Jacques Bernoulli in a discussion of probabilistic knowledge based on multiple differing observations.[100] Again, students of Locke will recognise that, for him too, knowledge of material things was probabilistic, and so they will be justified in concluding the period is more or less Lockean. But again, rather than focus on those aspects of scientific observation which were continuous with Locke and which persisted into the nineteenth century, I want to focus on aspects of the period which were intrinsic to the discourse of sensibility, specifically, on the continuity between scientific observation and aesthetic/moral theories. That is, here, the scientific genius was proximate to, perhaps even indistinguishable from, the artistic/moral genius.

Like for Ménuret, for Senebier, the idealised observer was characterised in terms of the particularity of genius: 'the observer must have [a quality of] Genius'.[101]

---

[93] Ménuret 1765, 312, 311.
[94] Ménuret 1765, 310.
[95] Ménuret 1765, 311.
[96] Ménuret 1765, 311.
[97] Senebier 1775, Vol. 1, 97.
[98] Senebier 1775, Vol. 1, 97, 131.
[99] Senebier 1775, Vol. 1, 223–224, citations on 223.
[100] Senebier 1775, Vol. 1, 230.
[101] Senebier 1775, Vol. 1, 13.

> Genius implies [possessing] all the qualities of the mind in their highest degree. [... The quality of] genius is thus that piercing gaze of the soul, which all at once grasps all the ideas relative to the object which absorbs it, which examines them separately, which first disentangles from them that which can enlighten it, and which, through this complete, swift and felicitous examination, soars towards sublime truths, and tears the sombre veil with which Nature confronts ordinary efforts. [... T]he man of genius has many more ideas than the man who lacks this quality [...]; he grasps a greater number of relationships.[102]

The idea that the observer-genius was the individual who noticed relationships/ *rapports* was prominent in the text and in the epistemology of the period.[103] Notwithstanding the section on analogy (understood here as 'relationships with more or less appreciable (*sensible*) resemblance'[104]), what actually constituted *rapports* was never made completely clear, with the idea relying on the primitive notion that they were simply known by the senses and noticed by the observer-genius. In fact, the text remained committed to the idea that the art of observation was not something which could be explicated completely, and despite what we might call the text's extended emphasis on 'rules for the direction of observation', in the end, for Senebier, there was something ineffable about the genius of observers, something which could only be learnt by intuition and by living inside the head of other geniuses:

> It is not easy for all men to establish the true relationships of distant Beings; it would thus be important for a large number [of them] to be able to follow the chain of ideas and observations which led the great Observers to those relationships. It would be necessary to inhabit the Mind (*Cerveau*) of a BONNET, a TREMBLEY, or a DE HALLER, to learn by intuition the art of observing Nature.[105]

Note that Senebier did not think of genius as being innate, and the text is aimed at the cultivation of the observer-genius.[106] With Helvétius and Rousseau, Senebier sees genius as being the product of cultivation, with the text focused extensively on questions of education.

Evidence of the continuity or complicity between aesthetics and natural philosophy which could be seen in this reliance on the ideas of genius and intuition was also found in Senebier's use of the descriptor '*peintre*' (literally, 'painter'). One of the skills of an observer-genius was the ability to represent or communicate

---

[102] '*Le génie suppose toutes les qualités de l'esprit à leur plus haut degré. [...] Le génie est donc cette vue perçante de l'ame, qui saisit tout d'un coup toutes les idées relatives à l'objet que l'occupe, qui les examine séparément, qui démêle d'abord au milieu d'elles ce qui peut l'éclairer, & qui par cet examen complet, prompt & heureux s'élance vers des vérités sublimes, & déchire le voile sombre que la Nature opposait à des efforts ordinaires. [... L]'homme de génie a beaucoup plus d'idées que celui qui en est privé [...]: il saisira un plus grand nombre de rapports*'. Senebier 1775, Vol. 1, 14–16.

[103] '*La science de l'Observateur n'est autre chose que la connaissance des rapports que les divers Etres ont entr'eux*'. (Senebier 1775, Vol. 1, 32.) See also Senebier 1775, Vol. 1, 93–94, 97, 137, 152, 155, Vol. 2, 42, 48–50; Singy 2006, 64–65.

[104] Senebier 1775, Vol. 2, 86.

[105] Senebier 1775, Vol. 2, 148.

[106] See the section on 'Des moyens de faire fleurir l'art d'observer' in Senebier 1775, Vol. 2, 146.

observations, and Senebier devoted an extended section of the text to 'l'Observateur Peintre de la Nature'.[107] There was ambiguity, in the period, over what was denoted by the noun *peintre*: it could mean a painter of pictures and a describer in language, either in prose or poetry (where today, the term tends to have aesthetic overtones and *décrire*/describe is more likely to be used when referring to the activities of a scientist).[108] This ambiguity was not a feature just of the word used by Senebier, but extended to the persona of the observer itself, as evidenced by the fifth and final section of the text, 'the art of observation [as] creator of the sciences and the arts'.[109]

> If Observation is the mother of the sciences, it is also that of the Arts; they all issue from Nature, whether we consider them relative to our pleasures, or as attending to our needs. Man creates nothing, he only combines the ideas he has received through the senses, or he reflects on the sensations he experiences, in order to draw new ideas from them. [...] We could define the Arts [as], *the means of grasping and employing those relationships that observation discovers between the Beings which comprise Nature, such that we can apply them most suitably to all that can bring pleasure or utility.*
>
> The aim of the Arts is fulfilled, when by an exact imitation of Nature, we have gratified the senses and moved the soul.[110]

The section included a specific discussion of the *beaux-arts*; the observer-geniuses here included Voltaire and, significantly, Richardson who 'not only grasps the great traits of a passion, he notices all its nuances, distinguishes all its characteristics'.[111] That there was a fundamental complicity between aesthetics and morality is a central feature of eighteenth-century moral sense theory. And so it is little surprising to find that moral sense is of interest to Senebier and that Francis Hutchinson and Adam Smith are the key references.[112] Moral science, too, is within the purview of the observer.[113]

The range of things which were unified under the rubric of the observer-genius was then extensive. The art of the observer covered speculative and experimental natural philosophy, natural theology, political and moral science, and the arts, including the fine arts. Much has been written about the novel of sensibility, and the relationship between vitalist medicine and this genre has long been noted.[114] Among other things, this chapter seeks to foreground what can be understood to be the proper epistemological functioning of the moral sense and so too, of the novel of sensibility. Under the influence of vitalist medicine, *sensibilité* became an epistemological term; within what we might call (tongues firmly in cheeks)

---

[107] Senebier 1775, Vol. 2, 1–36.

[108] '*PEINTRE. s.m. Celui qui fait profession de peindre. [...] Il se dit aussi De ceux qui représentent vivement les choses dont ils parlent, dont ils traitent, soit en Prose, soit en Poësie. Cet Orateur est un grand peintre. Ce Poëte est un excellent peintre.*' (Anonymous 1762.)

[109] Senebier 1775, Vol. 2, 161–321.

[110] Senebier 1775, Vol. 2, 279–280. Emphasis in original.

[111] Senebier 1775, Vol. 2, 281.

[112] Senebier 1775, Vol. 2, 201. See also Ménuret 1765, 311–312.

[113] Senebier 1775, Vol. 2, 205.

[114] See my introductory chapter for a more detailed discussion of this relationship. See also Vila 1998; Packham 2012.

'French empiricism', moral sense, and aesthetic and affective responses, including responses to literature, had the same epistemological status as the 'five senses'. The point, then, is this: there was no clear distinction in this period between someone who was an acute observer of physical phenomena, a doctor who felt or sensed a patient's fever, for example, or an observer of moral phenomena, a moralist who felt or sensed outrage at the plight of a beggar or, to invoke Diderot's 'Éloge De Richardson', the plight of Clarissa.[115]

9.3 Conclusion

There is much to be said about the persona of the philosopher, doctor, or moralist, the hero of sensibility, and the manner in which sensibility needed to be cared for and deliberately cultivated. This was the grounds of the period's interest in hygienic practices, about which much has been, and continues to be, written, including in chapters in this volume.

In speculative terms, sensibility was the unifying principle beneath a proliferation of differences. Considered as a discourse, 'Enlightenment vitalism' produced particulars, specifically particular subjects. I do not mean to speak of the production of particular subjects in a Foucauldian sense, that is, as the production of individuals who were subjected and normalised within a given discourse. Indeed, one of the most important features of the discourse of sensibility was its production of non-normal, but also non-deviant, always particular, subjects. Further, I have emphasised the difference between the embodied subject of sensibility and the universal rational subject, which was such a prominent feature in Locke and in Kant (to return to my introductory discussion of the 'epistemological paradigm'), and which has so often been taken as the preeminent feature of Enlightenment thought. This chapter has drawn out features of the period which may be thought to be paradigmatically 'Romantic'. That is, if the Enlightenment was associated with the pre-eminence of reason, then aspects of the period which have been the subject of this chapter must not, *ex hypothesi*, be of the period: they become 'Proto-Romantic'. It is to avoid this problematic term that Peter Hanns Reill has suggested that of 'Enlightenment vitalism'.[116]

I do not want to say that the subject of sensibility was conceptualised without universal intent; it was. As medical science, the discourse of sensibility did make universal claims and, to recall just one instance, Le Camus was quite specific about his belief in the homogeneity of the rational soul. Nor do I want to say that the subject of sensibility was irrational. But even if the period did not always conform to it, and even as the period struggled against it, often by invoking a rational and immortal soul, as a discourse, this universal aspiration operated to produce embodied

---

[115] Diderot 1762.
[116] Reill 2005. See also Packham 2012.

particularity and as a consequence, there was a distinct tendency to epistemological particularity and a cult of genius.

I have two things to say by way of conclusion. First, one of the problems of the 'epistemological paradigm' is that it presumes epistemology to be first philosophy. Yet 'theories of knowledge' did not emerge as a discrete philosophical discipline until the nineteenth century and 'epistemology' did not become their label until 1854.[117] Yet in this chapter, I *have* given precedence to the theory of knowledge; I have *not*, however, done this by relying on a notion of epistemology in a narrow or contemporary sense. In making clear the effects of the discourse of sensibility, I am seeking to revise general understandings of the epistemology of the French Enlightenment. In the thought of the period, the theory of knowledge was not *in isolation* foundational, but was situated within a much broader medico-philosophical anthropology. Second, the operation of the discourse of sensibility has implications for the practice of intellectual history, specifically for understandings of genre. The discourse ran together at least three genres which are today generally thought to be separate and which, aligned with the disciplinary boundaries in the contemporary university, are generally taken to be separate fields of scholarly inquiry: literature, rationalist metaphysics/speculative natural philosophy, and empirical science/ experimental natural philosophy and natural history. It was no accident that in this period, the philosophical novel was often the genre of choice, and it was no accident that novelists conceived of their task, the study of the truth of the human heart, of the affects of the heart, in terms identical to that in which doctors and moral philosophers conceived of theirs.

Acknowledgments I would particularly like to thank Alexander Cook, Peter Cryle, and Kim Hajek for their assistance in helping me preparing this chapter.

References

Anonymous. 1762. Peintre. In *Dictionnaire de l'Académie française*, 2 vols., vol. 2, 4th ed, 337. Paris: Chez la Vve B. Brunet.
Anonymous. 1765a. Observation. In *Encyclopédie ou Dictionnaire raisonné des arts et des métiers*, 35 vols., vol. 11, ed. Denis Diderot and Jean Le Rond D'Alembert, 313–321. Paris: Briasson, David, Le Breton & Durand.
Anonymous. 1765b. Plaisir (morale). In *Encyclopédie ou Dictionnaire raisonné des arts et des métiers*, 35 vols., vol. 12, ed. Denis Diderot and Jean Le Rond D'Alembert, 689–691. Paris: Briasson, David, Le Breton & Durand.
Anonymous. 1765c. Sens (Métaphysique). In *Encyclopédie ou Dictionnaire raisonné des arts et des métiers*, 35 vols., vol. 15, ed. Denis Diderot and Jean Le Rond D'Alembert, 24–27. Paris: Briasson, David, Le Breton & Durand.
Anonymous. 1765d. Sensations (Métaphysique). In *Encyclopédie ou Dictionnaire raisonné des arts et des métiers*, 35 vols., vol. 15, ed. Denis Diderot and Jean Le Rond D'Alembert, 34–38. Paris: Briasson, David, Le Breton & Durand.

---

[117] Haakonssen 2006.

Anstey, Peter R. 2005. Experimental versus speculative natural philosophy. In *The science of nature in the seventeenth century*, ed. P.R. Anstey and J.A. Schuster, 215–242. Dordrecht: Springer.

Boury, Dominique. 2008. Irritability and sensibility: Key concepts in assessing the medical doctrines of Haller and Bordeu. *Science in Context* 21(4): 521–535.

Cheung, Tobias. 2008. Regulating agents, functional interactions, and stimulus-reaction-schemes: The concept of 'Organism' in the organic system theories of Stahl, Bordeu, and Barthez. *Science in Context* 21(4): 495–519.

Condillac, Etienne Bonnot de. 1754/1970. Traité des sensations. In *Oeuvres complètes*, 1–327. Geneva: Slatkine Reprints.

Cunningham, Andrew. 2007. Hume's vitalism and its implications. *British Journal for the History of Philosophy* 15(1): 59–73.

Diderot, Denis. 1755. Epicuréisme ou Epicurisme. In *Encyclopédie ou Dictionnaire raisonné des arts et des métiers*, 35 vols., vol. 5, ed. Denis Diderot and Jean Le Rond D'Alembert, 779–785. Paris: Briasson, David, Le Breton & Durand.

Diderot, Denis. 1758/1875–1877. Réflexions sur le Livre de l'Esprit. In *Oeuvres complètes de Diderot*, 20 vols., vol. 2, ed. J. Assézat, 267–274. Paris: Garnier.

Diderot, Denis. 1762. Éloge de Richardson, auteur des romans de Paméla, de Clarisse et de Grandisson. In *Oeuvres complètes de Diderot*, 20 vols., vol. 5, ed. J. Assézat, 211–227. Paris: Garnier.

Dulieu, Luis. 1952. Les articles d'Henri Fouquet dans l'Encyclopédie. *Revue d'histoire des sciences et de leurs applications* 5(1): 18–25.

Fouquet, Henri. 1765. Sensibilité, Sentiment (Médecine). In *Encyclopédie ou Dictionnaire raisonné des arts et des métiers*, 35 vols., vol. 15, ed. Denis Diderot and Jean Le Rond D'Alembert, 38–52. Paris: Briasson, David, Le Breton & Durand.

Gaukroger, Stephen. 2010. *The collapse of mechanism and the rise of sensibility: Science and the shaping of modernity, 1680–1760*. Oxford: Oxford University Press.

Haakonssen, Knud. 2006. The history of eighteenth-century philosophy: History or philosophy? In *The Cambridge history of eighteenth-century philosophy*, ed. Knud Haakonssen, 3–25. Cambridge: Cambridge University Press.

Helvétius, Claude A. 1758. *De L'Esprit*. Paris: Durand.

Jaucourt, Louis de. 1765. Sens moral. In *Encyclopédie ou Dictionnaire raisonné des arts et des métiers*, 35 vols., vol. 15, ed. Denis Diderot and Jean Le Rond D'Alembert, 28–29. Paris: Briasson, David, Le Breton & Durand.

Kaitaro, Timo. 2008. Can matter mark the hours? Eighteenth-century vitalist materialism and functional properties. *Science in Context* 21(4): 581–592.

Knight, Isabel. 1968. *The geometric spirit: The Abbé de Condillac and the French enlightenment*. New Haven/London: Yale University Press.

Le Camus, Antoine. 1769. *Medecine de l'esprit*, 2nd ed, 2 vols. Paris: Ganeau.

Le Roy, Charles-Georges. 1765. Homme (Morale). In *Encyclopédie ou Dictionnaire raisonné des arts et des métiers*, 35 vols., vol. 8, ed. Denis Diderot and Jean Le Rond D'Alembert, 274–278. Paris: Briasson, David, Le Breton & Durand.

Legée, Georgette. 1991. La physiologie dans l'œuvre de Jean Senebier. *Gesnerus* 49(3–4): 307–322.

Locke, John. 1690/1849. *An essay concerning human understanding*, 30th ed. London: William Tegg & Co.

Marx, Jacques. 1974. L'art d'observer au XVIIIe siècle: Jean Senebier et Charles Bonnet. *Janus* 61: 201–220.

Ménuret de Chambaud, Jean-Joseph. 1765. Observateur. In *Encyclopédie ou Dictionnaire raisonné des arts et des métiers*, 35 vols., vol. 11, ed. Denis Diderot and Jean Le Rond D'Alembert, 310–313. Paris: Briasson, David, Le Breton & Durand.

Moravia, Sergio. 1978. From *Homme machine* to *Homme sensible*: Changing eighteenth-century models of man's image. *Journal of the History of Ideas* 39(1): 45–60.

Morris, David B. 1990. The Marquis de Sade and the discourses of pain: Literature and medicine at the revolution. In *The languages of psyche: Mind and body in enlightenment thought*, ed. G.S. Rousseau and Roy Porter, 291–330. Oxford/Los Angeles: University of California Press.

Packham, Catherine. 2012. *Eighteenth-century vitalism: Bodies, culture, politics*. Basingstoke: Palgrave Macmillan.

Reill, Peter Hanns. 2005. *Vitalizing nature in the enlightenment*. Berkeley: University of California Press.

Rey, Roselyne. 1995. Vitalism, disease and society. In *Medicine in the enlightenment*, ed. Roy Porter, 274–288. Amsterdam: Rodopi.

Rey, Roselyne. 2000. *Naissance et développement du vitalisme en France de la deuxième moitié du 18e siècle à la fin du Premier Empire*. Oxford: Voltaire Foundation.

Riskin, Jessica. 2002. *Science in the age of sensibility: The sentimental empiricists of the French enlightenment*. Chicago: University of Chicago Press.

Senebier, Jean. 1775. *L'art d'observer*, 2 vols. Geneva.

Singy, Patrick. 2006. Huber's eyes: The art of scientific observation before the emergence of positivism. *Representations* 95(1): 54–75.

Tipton, Ian. 1996. Locke: Knowledge and its limits. In *British philosophy and the age of enlightenment*, Routledge History of Philosophy, vol. 5, ed. Stuart Brown, 69–95. London/New York: Routledge.

Vila, Anne C. 1998. *Enlightenment and pathology. Sensibility in the literature and medicine of eighteenth-century France*. Baltimore: Johns Hopkins University Press.

Wolfe, Charles T. 2012. Forms of materialist embodiment. In *Anatomy and the organization of knowledge, 1500–1850*, ed. Matthew Landers and Brian Muñoz, 129–144. London: Pickering and Chatto.

Wolfe, Charles T., and Motoichi Terada. 2008. The animal economy as object and program in Montpellier vitalism. *Science in Context* 21(4): 537–579.

Yolton, John W. 1990. The way of ideas: A retrospective. *The Journal of Philosophy* 87(10): 510–516.

Yolton, John W. 1991. *Locke and French materialism*. Oxford: Oxford University Press.

Chapter 10
Sensibility in Ruins: Imagined Realities, Perception Machines, and the Problem of Experience in Modernity

Peter Otto

Abstract Sensibility is commonly thought to distinguish the democratic present from the authoritarian past, while at the same time, because it depends on a refinement not available to the crowd, setting it in opposition to the excesses of consumer culture. Through a discussion of Mathew Lewis's *The Monk* and of Coleridge's response to its publication, this paper sketches the outline of a different narrative, a destabilising counterpart of the first, in which sensibility, in its radical and conservative forms, leads to the ungrounded delights of the consumer culture from which it claims to turn. The paper begins with the role played by the camera obscura as an analogy for perception, which marks the point from which these contradictory narratives of sensibility arise and begin to diverge. It concludes by suggesting that for some readers of gothic fictions, such as the modern hedonists described by Campbell in *The Romantic Ethic and the Spirit of Modern Consumerism*, the interdependence of nature and custom (fiction) opens the possibility that culture could become a stage for the production of constantly changing pleasurable affect, for the fabrication of new desires, and for performing 'other' selves.

> everyday experience is based on three lines of separation: between 'true life' and its mechanical simulation; between the objective reality and our false (illusory) perception of it; between my fleeting affects, feelings, attitudes, and so on, and the remaining hard core of my Self. All these three boundaries are threatened today.[1]

Writing in the *Critical Review* for February 1797, Coleridge famously describes *The Monk* (1796) as 'a *mormo* for children, a poison for youth, and a provocative for the debauchee', notes that its author, Matthew Lewis, is 'a man of rank and fortune' who 'signs himself a LEGISLATOR!', and then frames both remarks with his own

---

[1] Žižek 1997, 133.

P. Otto (✉)
School of Culture and Communication, University of Melbourne,
3010 Parkville, VIC, Australia
e-mail: peterjo@unimelb.edu.au

breathless reaction—'We stare and tremble'.[2] These scenes of dangerous consumption, disgraceful creation, and horrified spectatorship rapidly eclipsed earlier, more positive reviews of Lewis's gothic fiction, while setting the stage for the adverse reviews that followed, which reached their emotional climax later that year in the fourth dialogue of Thomas Mathias's *The Pursuits of Literature*. Where Coleridge stared and trembled, Mathias is furious:

> A legislator in our own parliament, [...] an elected guardian and defender of the laws, the religion, and the good manners of the country, [who] has neither scrupled nor blushed to depict, and to publish to the world, the arts of lewd and systematic seduction, and to thrust upon the nation the most open and unqualified blasphemy against the very code and volume of our religion. And all this, with his name, style, and title, prefixed to the novel or romance called 'THE MONK'.[3]

Citing the charge of obscenity levelled against John Cleland's *The Memoirs of a Woman of Pleasure* (1748–1749), and the successful prosecution of Edmund Curll for the publication of *Venus in the Cloister* (1724), Mathias declares elements of the book to be '*actionable at Common Law*' and, raising the stakes again, classifies the book itself as 'a new species of legislative or state-parricide'.[4]

More recent accounts of *The Monk* often remain close to the schema established by Coleridge and Mathias, although inverting the values ascribed to its cardinal points: 'legislative or state-parricide' accordingly becomes the trope of emancipation from restraint, and the novel is thought to analyse 'repressed desire' (Norton), lift 'the lid on repressed sexuality' (Mishra), or defend 'the concept of individual desire and of the right to articulate that desire in both speech and action' (Jones).[5] *The Monk* is aligned with, on the one hand, the Enlightenment pursuit of happiness, culminating in the fall of the Bastille and, on the other hand, Freudian psychoanalysis: Ambrosio appears as a representative of the *Ancien Régime* and the death drive (Thanatos), Lorenzo and Raymond are cast as heroes of revolution and the sexual instinct (Eros), and Coleridge is left as anxious reactionary and judgemental superego.

This psychological/historical interpretation is compelling—so much so that, when one turns again to Lewis's novel, Coleridge's review, and Mathias's satiric poem, it is surprising to find they are more concerned by the easy *manufacture* and consequent widespread *proliferation* of desire than the emancipation or repression of a pre-existing drive, instinct, or subject. In *The Monk*, the secular 'customs' and 'fashions' of Madrid, coupled with the practices of 'ill religion' they superficially oppose, are blamed for these developments, which begin with the eclipse of the real by fictitious entities and imaginary worlds. But for Coleridge and Mathias, *The Monk* is itself a machine that manufactures a modern secular version of these

---

[2] Coleridge 1797, 197, 198.
[3] Mathias 1797, ii.
[4] Mathias 1797, ii–iii.
[5] Norton 2000, 117; Mishra 1994, 233; Jones 1990, 129. See also Howells 1995, 67; Huckvale 2010, 93; Krzywinska 2006, 72; Macdonald 2000, 79. The most important alternatives to this approach include Sedgwick 1981; Mulman 1998.

phenomena by drawing readers into an unreal world that rouses the 'unnatural' rather than the 'natural affections', creating a sense of false identity and false consciousness, and in so doing, threatens the capacity for rational judgement.

Much the same could be said about any number of gothic fictions; but in contrast to its compeers, which are 'manufactured' with 'little expense of thought or imagination', *The Monk* is, according to Coleridge and Mathias, 'the offspring of no common genius' and of no common member of society.[6] The former explains *The Monk*'s literary power and, when coupled with its subject-matter, its novelty, and it does this by drawing on the often-remarked conjunction in this book of literary genius and hack writer, compelling art and debased fiction, singularity and lack of originality. As described in an anonymous review, which appeared in *The European Magazine* in the same month as Coleridge's review was published, 'This singular composition, which has neither *originality*, *morals*, nor *probability* to recommend it, has excited, and will still continue to excite, the curiosity of the public. Such is the irresistible energy of genius'.[7]

This conjunction of opposites was, for many, sufficient to convict Lewis of bad taste and moral turpitude. But for Coleridge and Mathias, when the signature of a 'Member of Parliament' is added to the mix, a still more serious threat becomes visible, which they loosely correlate with the French revolution *and* a modern culture of information. As Mathias explains to his readers,

> We are no longer in an age of ignorance, and information is not partially distributed according to the ranks, and orders, and functions, and dignities of social life. [...] We no longer look exclusively for learned authors in the usual place, in the retreats of academic erudition, and in the seats of religion. Our peasantry now read *The Rights of man* on mountains, and moors, and by the way side; [...] Our *unsexed* female writers now instruct, or confuse, us and themselves in the labyrinth of politics, or turn us wild with Gallic frenzy.[8]

The disorder promoted by this oxymoronic conjunction of instruction and confusion, bewilderment and passion, learned author and peasant reader, gender and absence-of-gender, England and France, is extended by the proliferation of popular entertainments, exemplified by gothic fictions, which introduce into the actual world a multitude of unreal-realities that variously eclipse, double, supplement, or vie with reality. In Lyotard's gnomic summary: the dissolution of cultural authority is, in modernity, followed by the 'discovery of the "lack of reality", together with the invention of other realities'.[9]

These developments suggest that the fictions that structure society and regulate the passions of its members can be changed by other fictions. Thomas Paine claims in *Rights of Man* (1791), for example, that although it might seem to have 'burst forth like a creation from a chaos', the French Revolution 'is no more than the consequence of a mental revolution priorly existing in France. The mind of the nation had changed beforehand, and the new order of things has naturally followed the new

---

[6] Coleridge 1797, 194.

[7] R. 1797, 111.

[8] Mathias 1797, ii.

[9] '*du peu de réalité de la réalité, associée à l' invention d'autres réalités*'. Lyotard 1997, 25.

order of thoughts'.[10] And on the other side of politics, Mathias warns that 'LITERATURE, *well or ill conducted,* IS THE GREAT ENGINE *by which,* I am fully persuaded, ALL CIVILIZED STATES *must ultimately be supported or overthrown*'.[11] In this context, Coleridge and Mathias glimpse in *The Monk*'s collocation of popular fiction, literary genius, and legislative authority, the emergence of a new kind of art, one which extends in order to profit from the malleability of the real.

During the eighteenth century, the most commonly prescribed antidote to the loss of cultural authority and consequent proliferation of ungrounded realities was provided by sensibility, as framed by the account of perception elaborated in John Locke's *An Essay Concerning Human Understanding* (1690), and that of absolute space as the sensorium of God, as proposed by Isaac Newton in his *Opticks* (1704). The first identified the senses and therefore perception as the interface between subject and object, mind and world, while the second linked the world and therefore the ideas that world roused in bodies, to God.[12] But although described as natural and spontaneous, the simple ideas received from nature through the senses, and the natural affections these ideas aroused in the spectator, are framed by custom and opinion. Indeed, Locke describes custom as 'a greater power than Nature'; Shaftesbury writes of 'the Force of Custom and Education in opposition to Nature', while Hutcheson admits that '*Associations* of Ideas' can 'raise the Passions into an extravagant Degree, beyond the proportion of real Good in the Object'.[13] All three writers are nevertheless confident that our organs of internal and external sensation, given the right conditions, will develop in ways that give us the ability to distinguish between reality and fiction, truth and falsehood, virtue and vice. But this raises the question of what customs and opinions (what fictions) are likely best to foster the propensities with which we are born. In the culture inhabited by Coleridge, Mathias, and Paine, this question moves to the centre of political debate.

According to Mathias, 'the stability of government and the empire of good sense'[14] depend on the fictions of traditional society, which although naturalised by history and tradition, are threatened equally by the levelling impulses of revolutionary France, a nascent mass media, and the emergence of a consumer culture. From this point of view, the receptivity of sensibility and its consequent vulnerability to re-narration identify it as a threat. Indeed, for the writers of *The Anti-Jacobin*, like George Canning, sensibility had many years ago been taken from France, her native home, by Rousseau and taught a 'New Morality'.[15] Conversely, in *The Monk*, which lies at the other end of the political spectrum, 'good sense' is closely associated with revolutionary fictions centred on sensibility and the paternal family, which by arousing the natural affections, bring society into 'a just correspondence and symmetry with

---

[10] Paine 1791, 89.

[11] Mathias 1797, i.

[12] Locke 1690/1959, Vol. 1, 298; Newton 1718, 379.

[13] Locke 1690/1959, Vol. 1, 88; Cooper 1723, Vol. 2, 45; Hutcheson 1728, 63.

[14] Mathias 1798, 32.

[15] Canning 1801.

the order of the world'.[16] Despite their obvious differences, both narratives are conservative, in the sense that they prescribe a return to our supposedly natural affections and predispositions.

In contrast, for Coleridge, the boundary between truth and falsehood is more equivocal, the return to our natural affections is more problematic, and 'the empire of good sense' therefore seems much more evanescent. Sensibility accordingly appears variously as poison, patient, and cure (villain, victim, and hero/heroine). As I will argue in the following pages, the secret complicity between these incompatible roles is foregrounded by *The Monk*, and is the primary source of the horror roused in Coleridge by that book. It suggests to Coleridge that sensibility, and the natural affections it rouses, are structured rather than merely being framed or articulated by culture, and therefore that the truths of sensibility are always shadowed by falsehood, and still more disturbingly, the voice of honest indignation is accompanied by the nightmare of symbolic castration. Sensibility, one might say, is always haunted by its own ruin.

By grounding knowledge in individual experience, sensibility is commonly thought to distinguish the democratic present from the authoritarian past, while at the same time, because it depends on a refinement not available to the crowd, setting it in opposition to the excesses of consumer culture.[17] In the following pages, I want to use Coleridge's review of *The Monk*, and the nightmare it recounts, to sketch the outlines of a rather different narrative, a destabilising counterpart of the first, in which sensibility, in its radical and conservative forms, leads to the ungrounded delights of the consumer culture from which it claims to turn. Our argument begins with the role played by the camera obscura as an analogy for perception, which marks the point from which these contradictory narratives of sensibility arise and begin to diverge. It concludes by suggesting that for readers of gothic fictions less anxious than Coleridge, such as the modern hedonists described by Campbell in *The Romantic Ethic and the Spirit of Modern Consumerism*,[18] the interdependence of nature and custom (fiction) opens the possibility that culture could become a stage for the production of constantly changing pleasurable affect, for the fabrication of new desires, and for performing 'other' selves.

10.1 The Camera Obscura of Perception

A camera obscura comprises a darkened chamber, with a circular opening in one of its walls, usually filled with a convex lens through which light is able to enter, and a screen, placed inside the chamber, facing the aperture. When placed in daylight, this simple optical machine automatically paints or draws on its screen a detailed miniature of the external world, which was often described as being 'much more lively

---

[16] Burke 1790, 48.
[17] Perhaps the most important recent revisionary account of sensibility is Gaston 2010.
[18] Campbell 1987.

and distinct than the best finished drawings of the greatest artist',[19] or still more enthusiastically, as 'A new Creation! deckt with ev'ry Grace!/Form'd by [the camera obscura's] Pencil, in a Moment's Space!'[20]

From the seventeenth until at least the early nineteenth century, this device provided a powerful analogy for the mechanics of sight. As Descartes explains in *La Dioptrique* (1637), 'it is said that the [dark] room represents the eye; the hole, the pupil; the lens, the crystalline humour [...]; and the sheet [screen], the internal membrane, which is composed of the optic nerve-endings'.[21] And as Newton writes in *Opticks* (1704), once these elements are in place, 'the Pictures of Objects [are] lively painted' on the back of our eyes, 'And these Pictures, propagated by Motion along the Fibres of the Optick Nerves into the Brain, are the causes of Vision'.[22] According to this conceit, the eye is 'like a small world in another small world: a dark chamber of infinite art, and without which all the beauties of the world would be as nothing'.[23]

In Locke's *Essay*, this analogy is broadened to provide a model for the operation of the senses in general, and for the relation between the data they produce and the mind. '[E]xternal and internal sensations', Locke reports,

> are the only passages that I can find of knowledge to the understanding. These alone, as far as I can discover, are the windows by which light is let into this *dark room*. For methinks, the understanding is not much unlike a closet wholly shut from light, with only some little openings left, to let in external visible resemblances, or ideas of things without: would the pictures coming into such a dark room but stay there, and lie so orderly as to be found on occasion, it would very much resemble the understanding of a man, in reference to all objects of sight, and the ideas of them.[24]

The analogy is used by Locke to establish the boundaries necessary if experience (external and internal sensation) is to provide a reliable foundation for knowledge: it defines an inner space, unencumbered by innate ideas, which is set apart from the external world *and* the passions of the body. For those strands of sensational psychology and the literature of sensibility that take Locke as their point of departure, the camera obscura of perception accordingly becomes the locus of sensibility (rational feeling): it provides the window through which the world touches (without itself being touched by) the body, rousing its natural affections, *and* the site where 'an orderly projection of the world, of extended substance, is made available for

---

[19] Adams 1794, Vol. 2, 198, 199.

[20] Anonymous 1747, 4.

[21] '*Car ils disent que cette chambre représente l'œil; ce trou, la prunelle; ce verre, l'humeur cristalline, [...] et ce linge, la peau intérieure, qui est composée des extrémités du nerf optique*'. Descartes 1985, Vol. 2, 166.

[22] Newton 1718, 12.

[23] Scheuchzer, Johann Jakob. 1732–1737. *Physique sacrée; ou, Histoire-naturelle de la Bible*. Amsterdam. Quoted in Stafford and Terpak 2001, 145.

[24] Locke 1690/1959, Vol. 1, 211–212.

inspection by the mind'.[25] Yet already in Locke's *Essay*, perception raises dilemmas that suggest how equivocal this foundation will be.

In the camera obscura of perception, the 'immediate object[s] of perception, thought, or understanding' are ideas—'sensation or perceptions in our understandings'—rather than objects in the external world.[26] Indeed, Locke warns that most ideas of sensation are 'no more the likeness of something existing without us, than the names that stand for them are the likeness of our ideas'.[27] We can therefore be certain of the simple ideas we receive but not of the reality they appear to represent.[28] It follows, as David Hume (1711–1776) concludes, that the 'ultimate cause' of

> those *impressions*, which arise from the *senses*, [...] [is] in my opinion perfectly inexplicable by human reason, and'twill always be impossible to decide with certainty, whether they arise immediately from the object, or are produc'd by the creative power of the mind.[29]

In a camera obscura, one can, of course, always step out from its 'dark room' to check whether its virtual landscapes correspond with those outside. This possibility conditions the association of the camera obscura with realism. Yet we can't step outside the camera obscura of perception. Consequently, if our senses were to change, our view of the world would also change. Locke therefore concedes the possibility that 'in other mansions' of the creation, there may be other 'and different intelligent Beings, of whose Faculties [we have] as little Knowledge or Apprehension, as a Worm shut up in one Drawer of a Cabinet hath of the Senses or Understanding of a Man'.[30] In this admission, one can glimpse a nascent gothic and an embryonic romantic sensibility, which dissolves the unified world of traditional metaphysics into multiple realities: the 'dark room' that defines the camera obscura of the worm exists within the larger space of the cabinet (another term used by Locke to describe the 'dark room' of the mind),[31] which is placed, one presumes, within a room inside a house that exists alongside other houses, each with their own rooms, cabinets, drawers, and so on. The camera obscura of perception is here represented as a (contingent) perceptual environment that operates within, and is therefore in part articulated by, a range of architectural, social, and cultural environments, rather than 'a pre-given world of objective truth'.[32]

The gap between perception and object, and the dependence of perception on an optical apparatus, are not necessarily cause for alarm if, following Newton, one

---

[25] Crary 1992, 46.
[26] Locke 1690/1959, Vol. 1, 169.
[27] Locke 1690/1959, Vol. 1, 168.
[28] Locke 1690/1959, Vol. 2, 185–186.
[29] Hume 1739–1740/1985, 84. Locke admits that whether creations of the imagination, such as a centaur, 'can possibly exist or no, it is probable we do not know' (Locke 1690/1959, Vol. 1, 501).
[30] Locke 1690/1959, Vol. 1, 146.
[31] Locke 1690/1959, Vol. 1, 48.
[32] Crary 1992, 39–40. This paragraph is drawn from my 'Inside the Imagination-machines of Gothic Fiction: Estrangement, Transport, Affect', which can be read as a companion-piece to the present essay (Otto 2011 b).

believes that absolute space is the sensorium of God and, following Locke, that our perceptual apparatus has been installed by God. Addison, for example, writing in *The Spectator* for June 1712, readily admits that objects excite ideas, such as light and colour, that 'are different from anything that exists in the objects themselves', and that we are therefore

> at present delightfully lost and bewildered in a pleasing Delusion, and we walk about like the Enchanted Hero of a Romance, who sees beautiful Castles, Woods, and Meadows; and at the same time hears the warbling of Birds, and the purling of streams.[33]

But for Addison, this 'pleasing Delusion' is a fiction designed by God so that we are unable 'to behold his Works with Coldness and Indifference'.[34]

Even in this formulation, sensibility fills the body with sensations that link subject and object only if both have a common origin in God. But as this belief falters, sensibility becomes the site of a converse movement, which fills the body with sensations shaped by circumstance, perceptual machines, and the desires they produce. This brings us back to *The Monk*, Coleridge's review of that book, and to the modern world inhabited by their readers, in which the demand for unreal-realities, and the 'new' experiences they prompt, threaten to eclipse the less protean pleasures offered by reality.

10.2 Preternatural Worlds and Unnatural Passions

'The horrible and the preternatural' are 'powerful stimulants', Coleridge writes, that seize 'on the popular taste, at the rise and decline of literature', when they minister respectively to 'the torpor of an unawakened [and] the languor of an exhausted, appetite'.[35] In Germany, gothic fictions awaken the appetites of an unsophisticated people and so make it possible for a national literature to emerge. But in England, the prognosis is less optimistic: gothic fictions are there a sign of the 'exhausted' appetites of a sophisticated people, and they mark the desire for desire (any desire), no matter how trivial or coarse the object that arouses it.

Although gloomy, the prognosis is not hopeless. Because gothic fictions are manufactured with 'little expense of thought or imagination', it can be hoped that a return of the disease (an exhausted appetite) will defeat the cure (a cheap stimulant): the public will eventually become 'wearied with […] shrieks, murders, and subterranean dungeons', and 'satiety will banish what good sense should have prevented'.[36] But this cure is unlikely to be effective against *The Monk*, a gothic fiction which is

---

[33] Addison 1712, 82.
[34] Addison 1712, 81.
[35] Coleridge 1797, 194.
[36] Coleridge 1797, 194.

also a work of genius, is created rather than manufactured, and everywhere 'discovers an imagination rich, powerful, and fervid'.[37]

In *Elements of Criticism* (1785), Lord Kames distinguishes, in terms of their effect on the passions, between a 'cursory narrative [...] of feigned incidents' and a work of genius, such as Shakespeare's *King Lear*.[38] The former leaves impressions on the mind which are 'faint, obscure, and imperfect'. The waking dream it conjures is therefore soon broken, reflection begins, and the 'slight pleasure' the story offers is 'counterbalanced by the disgust it inspires for want of truth'. In contrast, the latter conjures appearances (ideal presence) which are indistinguishable from real presence, whether the subject is 'fable or a true history'. In this vividly imagined virtual reality, we perceive every object in our sight, and the mind, totally occupied with an interesting event, 'finds no leisure for reflection'.[39] While in this state, we are therefore 'susceptible of the strongest impressions'.[40]

Kames adds that a 'chain of imagined incidents linked together according to the order of nature, finds easy admittance into the mind', and conversely, that we admit only with 'great difficulty' imagined facts—'our judgement revolts against an improbable incident; and if we once begin to doubt of its reality, [...] it will require more than an ordinary effort, to restore the waking dream'.[41] He is therefore able to admit with equanimity that the 'extensive influence which language hath over the heart' rests on the slight foundation offered by 'ideal presence'—as conjured by memory, literature, history, and so on. This influence, he continues,

> more than any other means, strengthens the bond of society, and attracts individuals from their private system to perform acts of generosity and benevolence. [...] Nor is the influence of language, by means of ideal presence, confined to the heart: it reacheth also the understanding, and contributes to belief.[42]

But in *The Monk*, Lewis uses his genius and imagination to create a waking dream at odds with reality, an apparently real unreality, in which scene, event, and character are all preternatural. First 'events are levelled into one common mass, and become almost equally probable, [so that] the order of nature may be changed'.[43] Next,

> the abominations which he pourtrays with no hurrying pencil, are such as the observation of character by no means demanded, such as 'no observation of character can justify, because no good man would willingly suffer them to pass, however transiently, through his mind'.[44]

---

[37] Coleridge 1797, 194.
[38] Home 1785, Vol. 1, 95.
[39] All Home 1785, Vol. 1, 95.
[40] Home 1785, Vol. 1, 101.
[41] Home 1785, Vol. 1, 102.
[42] Home 1785, Vol. 1, 100.
[43] Coleridge 1797, 194.
[44] Coleridge 1797, 195.

Further, in this unreal world, the Monk acts 'under the influence of an appetite which could not co-exist with his other emotions'.[45] Far from being admitted with great difficulty, these irregular imagined-facts, linked in ways contrary to the course of nature, gain 'easy admittance into the mind' of readers who are happy to suspend disbelief.

In *Biographia Literaria*, Coleridge claims that the entire '*materiel* and imagery' of gothic fiction's waking dreams are

> supplied *ab extra* by a sort of mental *camera obscura* manufactured at the printing office, which *pro tempore* fixes, reflects and transmits the moving phantasms of one man's delirium, so as to people the barrenness of an hundred other brains.[46]

This remarkable imagination machine not only multiplies phantasms, annexes the mind to itself (as the screen/auditorium within which its phantoms come to life), but also, through the passions they arouse in readers, gives these phantasms purchase on real bodies and the real world. This is bad enough. But *The Monk*, because it is a work of genius, exerts a still more powerful hold on readers and their passions: it adds 'subtlety to a poison by the elegance of its preparation'.[47] When this intoxicating mix is authorised by the imprimatur of a Member of Parliament, we do indeed have, at least from Coleridge's point of view, 'a *mormo* for children, a poison for youth, and a provocative for the debauchee'.[48]

10.3 Narrative and Desire

To this point in the review it would be reasonable to complain that Coleridge's animus against *The Monk* is informed by cultural phenomena that the novel itself seems to criticise. Lewis's Madrid is awash with desire which, rather than rising fully formed from within the body, as the humoral theory of the passions would suggest, is produced by a multitude of textual, cultural, or perceptual objects and machines. According to Locke, desire is produced when there is a discrepancy between our '"idea of delight" and the object whose possession would provide the sensation of delight'.[49] And desire, understood in this way, is sufficient to bring all the passions into being, from joy and sorrow, to hope, fear, despair, envy, anger, and shame.[50] But in the ungrounded world of Lewis's Madrid (and one might add of Coleridge's London), this kind of disjunction is so common, and desire is accordingly so easily produced, that passion often seems comical: a neck glimpsed beneath a thick veil, a tear falling on a cheek, a half-exposed breast, romances read by a

---

[45] Coleridge 1797, 196.
[46] Coleridge 1983, Vol. 1, 48, Note.
[47] Coleridge 1797, 196.
[48] Coleridge 1797, 197.
[49] Armstrong and Tennenhouse 2006, 138.
[50] Locke 1690/1959, Vol. 1, 304.

young man to an older woman, even a child's unsupervised reading of the Bible, and so on are all productive of desire, often of overwhelming desire, able to muffle the voice of reason.[51]

In pre-revolutionary Madrid, one might expect the Church to exert power through a narrative that trumps the mundane differential between pleasure and pain with the eternal disjunction between heaven and hell. In the first chapter of the book, Ambrosio uses this narrative and his skills as an orator to turn briefly the desires of his audience from earthly to heavenly things. While he describes hell, 'Every Hearer looked back upon his past offences, and trembled: The Thunder seemed to roll, whose bolt was destined to crush him, and the abyss of eternal destruction to open before his feet', and when he turns his attention to Heaven, 'They were transported to those happy regions which He painted to their imaginations in colours so brilliant and glowing'.[52] But in Madrid, a city teeming with imaginary worlds and fictitious objects, religious visions are soon lost amongst a host of secular ones.

An apparent alternative to religious and conventional-secular narratives is first offered by the story of Agnes and Raymond. When Donna Rodolpha refuses to allow them to marry, they decide to elope. It is *their* narrative, represented by this decision, rather than narratives provided by the family, social convention, or the Church, that here establishes the differential between present pain and future pleasure. But they quickly fall victim to much more powerful narratives, represented most vividly by the bleeding nun, a ghost whose appearance on the scene, as in any good gothic story, chains the auditor/reader's (Raymond's) 'limbs […] in second infancy', while transporting him (in imagination) into a preternatural world.[53] Like the phantasms described by Coleridge, this visionary figure is an *ab intra* delusion which, because she presents as an *ab extra* reality, is able to draw Raymond into her world, and in so doing, claim his body as her own. As she announces 'in a low sepulchral voice':

> Raymond! Raymond! Thou art mine!
> Raymond! Raymond! I am thine!
> In thy veins while blood shall roll,
> I am thine!
> Thou are mine!
> Mine thy body! Mine thy soul!—[54]

Even when the bleeding nun leaves him, and the 'blood which had been frozen' in his veins rushes 'back to [his] heart with violence',[55] she continues to haunt his fancy, and each night, returns to act the same scene once again. It is only the Wandering Jew, an ideal presence still more powerful than the nun, who is able to free Raymond. He does this by unveiling a spectacle—'a burning Cross impressed

---

[51] Lewis 1796/1973, 9, 79, 65, 134–135, 259.
[52] Lewis 1796/1973, 19.
[53] Lewis 1796/1973, 162.
[54] Lewis 1796/1973, 160.
[55] Lewis 1796/1973, 161.

upon his brow'[56]—which, by drawing the bleeding nun into the narrative that structures his life, enables her own story to be retold.

Even 'truth' is unable to escape this economy. Towards the end of *The Monk*, when the mother St. Ursula unveils the gap between the appearance and reality of monastic life, she produces a desire for revenge that leads to some of the most violent scenes in the book. In their shocking climax, the hell previously conjured by Ambrosio seems to have become real:

> Some employed themselves in searching out the Nuns, Others in pulling down parts of the Convent, and Others again in setting fire to the pictures and valuable furniture, which it contained. [...] The flames, rising from the burning piles caught part of the Building, which being old and dry, the conflagration spread with rapidity from room to room. The Walls were soon shaken by the devouring element: The Columns gave way: The Roofs came tumbling down upon the Rioters, and crushed many of them beneath their weight. Nothing was to be heard but shrieks and groans; The Convent was wrapped in flames, and the whole presented a scene of devastation and horror.[57]

The contrast between this secular hell (which brings the *ancien regime* to its catastrophic end) and the heaven of the paternal family (anticipated in the marriage of Raymond and Agnes, and of Lorenzo and Virginia) provides an eloquent instance of the mismatch between a past that still casts its shadow on the present and a utopian future which seems almost within reach, which is constitutive of many modern forms of desire.

It is possible, of course, to believe that the paternal family really is a secular heaven, within which the gap between our 'idea of delight' and the object 'whose possession would provide the sensation of delight' has been narrowed. As the narrator remarks, 'if ever [Raymond and Agnes, and Lorenzo and Virginia] felt Affliction's casual gales, they seemed to them gentle as Zephyrs, which breathe over summer-seas'.[58] But this utopian conclusion is balanced by its dystopian contrary, seen in the fate of Ambrosio, who at the end of the novel, is no longer able to propel himself towards the object of desire (oblivion) that he is still able to imagine:

> The Eagles of the rock tore his flesh piecemeal, and dug out his eye-balls with their crooked beaks. A burning thirst tormented him; He heard the river's murmur as it rolled beside him, but strove in vain to drag himself towards the sound.[59]

On the seventh day, completing the deconstruction of the world created by God, Ambrosio is swept into the river and so, one imagines, into Hell. But the opposition between Heaven and Hell is here the archetype for a life animated not by a gap between earthly and transcendental realities, but by the mismatch between our endless proliferating ideas of delight and the mundane objects within reach.

---

[56] Lewis 1796/1973, 172.
[57] Lewis 1796/1973, 357–358.
[58] Lewis 1796/1973, 420.
[59] Lewis 1796/1973, 442.

10.4 The Mutability of Images

Lewis is so acutely aware of the ways in which fiction generates desire that it is hardly surprising to find in *The Monk* an allegory of the production and consumption of gothic fictions that rivals the one sketched by Coleridge. Rather than a 'mental *camera obscura* manufactured at the printing office', Lewis's reflecting/projecting machine is 'a mirror of polished steel, the borders of which were marked with various strange and unknown characters'.[60] As Matilda explains to Ambrosio, when certain words are spoken, 'the Person appears in it on whom the Observer's thoughts are bent'.[61] Ambrosio has recently returned to his Cell after attempting to rape Antonia. When he takes the mirror and Matilda pronounces the magic words, she is therefore the person who is seen:

> a thick smoke rose from the characters traced upon the borders, and spread itself over the surface. It dispersed again gradually; A confused mixture of colours and images presented themselves to the Friar's eyes, which at length arranging themselves in their proper places, He beheld in miniature Antonia's lovely form.[62]

The characters traced on the mirror's margins imply that it is a book, a metaphor for the gothic fiction that we are reading, and the mix of colours and images that eventually take Antonia's form recall the figures of real *and* ideal presence. As a mirror, and like a camera obscura, this device offers a window on the real world: it carries into Ambrosio's Cell an image or miniature (an 'external visible resemblance'[63]) of that reality. But unlike a conventional mirror or camera obscura, it is not confined to objects in its vicinity and can be focused by the observer's desire and curiosity: it selects from the world the 'objects' that the observer wants to see. As this suggests, the mirror is also a device for projecting Ambrosio's *ab intra* delusions as if they were *ab extra* projections of reality. This makes Ambrosio both spectator (of an image drawn from the external world) and co-author with Matilda (through selection) of what he sees.

The miniature re-presents a figure of sensibility, the most prominent in Lewis's book, which even at this point in the narrative, might be thought to turn our gaze out to the real world and its Creator. According to one strand of Christian thought, 'the voluptuous quality' of sacred images 'refers to the love that should be directed toward the divinity'.[64] Yet an image is also 'a body, a material presence', and the voluptuous is experienced when that body brushes up against our own.[65] Seen in this light, as a second strand of Christian thought contends, images lead us away from God, into the world of the flesh. In the pages of *The Monk*, Matilda personifies this demonic power of the image. Like the demons described by Tertullian, she haunts

---

[60] Lewis 1796/1973, 270.
[61] Lewis 1796/1973, 270.
[62] Lewis 1796/1973, 271.
[63] Locke 1690/1959, Vol. 1, 212.
[64] Goodrich 1995, 57.
[65] Goodrich 1995, 57. For a useful discussion of religious imagery in *The Monk* see Mulman 1998.

images of the sacred (of nature, a hermit's retreat, the hermit's pious meditation, the Virgin Mary) in order to clear the path, offered by these same images, from the soul to the body, and from truth to falsehood (nature is false; the hermit's retreat is a pretence; his meditation a lie; even the portrait of the Virgin Mary was drawn by an artist who took Matilda as his model). The scene of reading/viewing we are considering moves just as decisively away from truth, taking the reader deep into the world of the false.

As Ambrosio continues to gaze into the mirror, the space within which Antonia is standing comes into focus, while she, gaining three-dimensional form and sensuous detail, becomes a moving figure on a private stage:

> The scene was a small closet belonging to [Antonia's] apartment. She was undressing to bathe herself. The long tresses of her hair were already bound up. The amorous Monk had full opportunity to observe the voluptuous contours and admirable symmetry of her person. She threw off her last garment, and advancing to the Bath prepared for her, She put her foot into the water. It struck cold, and She drew it back again. Though unconscious of being observed, an inbred sense of modesty induced her to veil her charms; and She stood hesitating upon the brink, in the attitude of the Venus de Medicis. At this moment a tame Linnet flew towards her, nestled its head between her breasts, and nibbled them in wanton play. The smiling Antonia strove in vain to shake off the Bird, and at length raised her hands to drive it from its delightful harbour.[66]

As one reads this passage, it is still possible to glimpse the *celestial* form represented by Antonia, in mention, for instance, of her 'long tresses [...] already bound up', or the 'voluptuous contours and admirable symmetry of her person'. Further, her 'inbred sense of modesty' brings the action momentarily to a standstill, as we see her hesitate on 'the brink' of the Bath, in 'the attitude of the Venus de Medicis'. The classical allusion emphasises that, in this moment, Antonia unites 'perfect beauty of form' and 'gracefulness of attitude', as well as 'the idea of conscious beauty' and of 'modesty',[67] in a form that evokes the eternal rather than the temporal world. As James Ussher writes, there

> is a kind of description, which [...] while it preserves nature, sometimes in a fine flight of fancy, throws an ideal splendor over the figures that never existed in real life. Such is [...] the inexpressible beauties that dwell upon the Venus of Medici, and seem to shed an illumination around her. [...] [T]he imagination carries the ideas of the beautiful [...] far beyond visible nature.[68]

But in the passage we are considering, this ideal form makes visible the rapidly widening gulf that divides Ambrosio from it, while throwing into relief Antonia's bodily form.

In fact, the scene we are watching runs in reverse the drama implied by the Medici Venus: rather than an ideal form emerging from the sea (metaphorically the chaos of the flesh), Antonia is about to step into her Bath, and as such, veiling is (for Ambrosio) a teasing prelude to the unveiling of her charms. The attitude assumed

---

[66] Lewis 1796/1973, 271.
[67] Beattie 1783, Vol. 1, 148.
[68] Ussher 1769, 48–49.

by the Medici Venus, adopted in natural response to the presence of an unwanted spectator, is an expression of her modesty, and of an inner nature consonant with the eternal order of things. But Antonia's modesty is described as 'inbred', a product of custom, while the impulses of nature are represented by the 'tame Linnet' whose 'wanton play' induces her to unveil 'her charms'. The form she now displays is more nearly that of a pagan, rather than Christian, goddess of love.

The conclusion is predictable: 'Ambrosio could bear no more: His desires were worked up to phrenzy. "I yield!" He cried, dashing the mirror upon the ground: "Matilda, I follow you! Do with me what you will!"'.[69] The line traced by Addison, from 'pleasing Delusion' (whether prompted by real or ideal presence), to the objects that excite these ideas (even though the latter 'are different from anything that exists in the objects themselves'),[70] and then to God is here eclipsed by a 'pleasing Delusion' that has cut itself loose from its model and so become mutable. Just how mutable it is can be gauged if one takes the mirror as a metaphor for the various optical environments within which Ambrosio is immersed (cell, garden, grotto, crypt, bedroom, and so on), each of which is in part animated by Matilda, first as the boy Rosario, then, in slow procession as the pure woman, wanton woman, willing servant of Ambrosio, masculine woman, master of Ambrosio, and finally, devil in the service of Satan. As Matilda changes, along with the perceptual environments she inhabits, Ambrosio's passions change as well, from pride in his purity and status, which is accompanied by an only partially sublimated sexual desire for the Virgin Mary, to a nascent homosexual and pederastic affection for the boy Rosario, then to a gradually increasing heterosexual desire for Matilda, followed by distaste for the masculine Matilda, an unconsciously incestuous lust for Antonia that is laced with necrophilia, an unconscious desire for his mother (Elvira), and finally to shame, guilt, torment, and ultimately despair as he measures his distance from the 'idea of delight' represented by God.

Notwithstanding the parallels between Coleridge's mental camera obscura and Lewis's magic mirror, this episode brings us to those aspects of *The Monk* that provoke not just Coleridge's distaste, but his anxiety as well. Indeed, one might say that in moments like this, Lewis makes it difficult for a reader like Coleridge to turn away.

10.5 Gothic Doubles

As we have seen, Coleridge hopes that gothic fictions, in England at least, will be defeated by the disease they are used to cure, leaving, one presumes, the body's active powers intact, ready for some future resurrection. The same cure is, however, helpless against *The Monk* because it turns what should be the real cure into a poison. This operates on a number of levels: genius is used against rather than in the

---

[69] Lewis 1796/1973, 271.
[70] Addison 1712, 81, 82.

service of nature; imagination is used to create an unreal world and ideal presence to convey abominations; the trembling innocence and sensibility of Antonia become the vehicle of the most voluptuous images; and so on. In this regard, *The Monk* is most ambitious in its representations of the Bible as likely to arouse sexual desire, and Coleridge's disgust accordingly reaches fever pitch:

> The impiety of this falsehood can be equalled only by its impudence. This is indeed as if a Corinthian harlot, clad from head to foot in the transparent thinness of a Cöan vest, should affect to view with prudish horror the naked knee of a Spartan matron![71]

And yet, following this outburst, he identifies a chapter in the Bible and, what is more, lines spoken by God, that might lend 'a *shadow* of plausibility to the *weakest* of these expressions'. And he adds that, in his view, 'it is not absolutely impossible that a mind may be so deeply depraved by the habit of reading lewd and voluptuous tales, as to use even the Bible in conjuring up the spirit of uncleanness'.[72]

Coleridge's emotion reflects his disgust at *The Monk*'s immorality and blasphemy, but it also registers his sense of how rapidly perception, and therefore thought and the will, can be rearticulated by bodily or cultural systems. Redirected from their proper, consciously intended objects, these elements of the psyche seem to become the vehicles of a second self, a gothic double that exerts on the conscious mind the metamorphic power of the false. In a letter to Thomas Poole, dated 5 November 1796, this double is generated by bodily pain: 'On Friday it only *niggled*', Coleridge writes,

> as if the Chief had departed from a conquered place, and merely left a small garrison behind [...] But *this morning* he returned in full force, and his name is Legion. Giant-fiend of a hundred hands, with a shower of arrowy death-pangs he transpierced me, and then he became a Wolf and lay a-gnawing at my bones![73]

In 'The Destiny of Nations', written in February 1796, and 'The Rime of the Ancient Mariner', begun in November 1797, the canvas is much broader: life itself is transpierced by its gothic double, respectively a 'nameless female' and the 'Nightmare *Life-in-Death*'. And in Coleridge's 'Christabel', also begun in 1797, Christabel's voice is usurped by her uncanny double, Geraldine.[74] In Coleridge's poetry, the agon between opposites and the trauma it prompts reach their climax in 'The Pains of Sleep', written in 1803, where Coleridge is racked by emotions that properly belong to his double (a figure who recalls the villains of gothic fictions, such as Ambrosio):

> Such punishments, I said, were due
> To natures deepliest stained with sin,—

---

[71] Coleridge 1797, 198.

[72] Coleridge 1797, 198.

[73] Coleridge 1895, Vol. 1, 174–175.

[74] Coleridge 1848, 245–267. Swann notes that 'men of letters reacted hysterically to *Christabel* because they saw the fantastic exchanges of Geraldine and Christabel as dramatising a range of problematically invested literary relations, including those between writers and other writers, and among authors, readers, and books'. What is at stake, she continues, 'is the identity and autonomy of the subject in relation to cultural forms'. (Swann 1985, 398.)

For aye entempesting anew
The unfathomable hell within,
The horror of their deeds to view,
To know and loathe, yet wish and do!
Such griefs with such men well agree,
But wherefore, wherefore fall on me?[75]

'I feel strongly and I think strongly', Coleridge writes to the radical John Thelwall in 1796, 'but I seldom feel without thinking or think without feeling. Hence, though my poetry has in general a hue of tenderness or passion over it, yet it seldom exhibits unmixed and simple tenderness or passion'. And conversely, he continues, 'My philosophical opinions are blended with or deduced from my feelings'.[76] This brings us back once more to sensibility and its two narratives, the first leading to nature and so to God, and the second to the ungrounded delights—falsehood or evil in Coleridge's lexicon—of a consumer culture.

In the year after completing his review of *The Monk*, Coleridge promises to 'devote [himself] to such works as encroach not on the anti-social passions—in poetry, to elevate the imagination and set the affections in right tune by the beauty of the inanimate impregnated as with a living soul by the presence of life'.[77] Although he couches it in romantic rather than materialist terms, Coleridge, like Newton, Locke, and Addison, assumes here that the data of sensation is a text, with a referent (the real), signatory (God), and addressee (the subject), that despite the limitations of his own vision, will lead us back ultimately to God. These lines are often taken as an anticipation of his life's work, but they also explain why Coleridge would judge his life's work a failure. To the extent that the referent, signatory, and addressee of the 'pleasing Delusion' of perception are not fully determinate realities, perception becomes what one might call, following Derrida, '*un objet littéraire*', subject to rearticulation by any number of textual, cultural, or perceptual objects and machines.[78] The data that, according to Coleridge, leads us to the true, paradoxically also takes us to the false; his true self is therefore always shadowed by its double.

On 13 October 1800, Coleridge published in the *Morning Post* a poem entitled 'The Mad Monk: An Ode in Mrs. Ratcliff's Manner', which represents this agon as a species of double voicing, in which each voice struggles to re-narrate the other.[79] Although the words heard by the narrator seem to come from a Monk or hermit (who remains unseen), they emerge from 'a cavern's mouth' on 'Etna's side'. Human and inhuman, heavenly and hellish voices therefore overlap, to the extent that each might be described as ventriloquist for the other: the human is heard through the inhuman; the inhuman speaks with the voice of the human. The same phenomenon appears in the poem as a whole, where the Monk's bipolar voice

---

[75] Coleridge 1848, 218.
[76] Coleridge 1895, Vol. 1, 197.
[77] Coleridge 1895, Vol. 1, 243.
[78] Derrida 2008, 131.
[79] Coleridge 1800, 3.

emerges from inside the voice of the narrator, which emerges from lifeless words on the page, a textual equivalent, perhaps, of the 'cavern's mouth' on 'Etna's side'.[80]

The divided voice constructed by these disparate sites of enunciation expresses a character and describes a world similarly divided into opposites held in each other's unwilling embrace. The speaker's love for Rosa (a 'maid divine'), when rearticulated by circumstance (a rival appears on the scene), becomes jealousy, which leads to her murder: '*I struck the wound*—this hand of mine!'[81] In the bipolar world that the Monk subsequently inhabits, 'The plot of mossy ground,/On which' he and Rosa once sat is also 'The roof of Rosa's grave!'[82]; amongst the flowers which cover the hills are those which recall the colour of Rosa's blood, turning a scene of beauty into one of remorse; and the 'red gleam' from the 'stormy clouds above', reflected on the 'downward trickling stream',[83] recalls the promise of forgiveness (Christ's blood) and his own despair. As the poem makes clear, on each of these levels, the demon is 'resemblance' rather than representation; it provides a path along which the elements of one narrative can rapidly be rearticulated in terms of another. Echoing Coleridge's decision 2 years earlier to turn from the 'anti-social passions', in the last lines of the poem, the narrator turns away from the 'Mad Monk'. In 'deep dismay', he writes, 'Down thro' the forest I pursu'd my way'.[84]

Just how difficult it is simply to turn away is suggested by Coleridge's discovery 2 years later that the daughter of Mary Robinson, the famous actor, writer, and courtesan, was planning to publish 'The Mad Monk' in a volume of poetry dedicated to her mother, which included poems by Matthew Lewis. Articulated in this way, even Coleridge's turn from the anti-social passions opens a road back to them. As he writes in horror to Robinson:

> —but I have a wife, I have sons, I have an infant Daughter—what excuse could I offer to my own conscience if by suffering my own name to be connected with those of Mr. Lewis […] I was the *occasion* of their reading the Monk […] Should I not be an infamous Pander to the Devil in the Seduction of my own offspring?—My head turns giddy, my heart sickens, at the very thought of seeing such books in the hands of a child of mine […] The mischief of these misery-making writings *laughs* at all calculations. On my own account therefore I must in the most emphatic manner decline all such connection.[85]

In order to manage the remarkable metamorphic power of modernity's cultural, perceptual, and imagination machines, *The Monk* turns to the paternal family. Within its spaces, quarantined from religious and secular/commercial narratives, women are free to be articulated by a new narrative, one that (supposedly) returns them to themselves: nuns accordingly become wives, mothers, and figures of sensibility—stable points of reference in an otherwise ungrounded world. But for others, the metamorphic power of modernity represents an opportunity. For Lewis himself,

---

[80] Coleridge 1800, 3, Lines 1–2.
[81] Coleridge 1800, 3, Line 33.
[82] Coleridge 1800, 3, Lines 27–28, 24.
[83] Coleridge 1800, 3, Lines 39, 38, 40.
[84] Coleridge 1800, 3, Lines 46–47.
[85] Coleridge 1932, Vol. 1, 234.

these metamorphic powers enable him to become one of the first literary celebrities. In his Temple of Health and Hymen, the charlatan James Graham constructs an environment in which, as he writes in *Il Convito Amoroso*, spectators are encouraged 'to transform themselves into what they love!—Yes, to absorb and assimilate the soul and body of their beloved—and to mix and intimately blend their substance with that of the object of their passion'.[86] In another version of the phenomenon, William Beckford constructs perceptual environments, most famously at Fonthill Abbey, designed for the production of constantly changing pleasurable affect.[87] More broadly, in gothic fiction, culture becomes a stage for experiencing unreal-realities and the passions they arouse.

But these developments are, for Coleridge, sure signs of the double he wants to dispatch, and to do so, he can turn only to his own weak will, which he must exert against a powerful adversary. As he writes in 1803:

> As he who passes over a bridge of slippery uneven Stones placed at unequal distances, at the foot of an enormous waterfall, is lost, if he suffers his Soul to be whirled away by its diffused every where nowhereness of Sound/but must condense his Life to the one anxiety of not Slipping, so will Virtue in certain Whirlwinds of Temptations.[88]

Against the loss of order implied by the 'every where nowhereness of sound', and to avoid being articulated by the inchoate world it represents, virtue must give itself entirely to a narrative ('the bridge of slippery uneven Stones') that is itself almost submerged by the waterfall.

Coleridge's late thought, as developed in *On the Constitution of the Church and State* (1830), attempts to strengthen the path of virtue by quarantining it from the amorphous modern world around it, as its key words suggest—cultivation, clerisy, constitution, state, national church. The opposition between high and low culture that this implies, informs the nineteenth-century tradition of liberal humanism; it also brings into sight humanism's inauthentic double—the modern world enabled by 'Roads, canals, machinery, the press, the periodical and daily press, the might of public opinion'.[89] In so doing, it ensures that humanism, like the discourse of sensibility from which it emerges, will always be haunted by its own ruin.

References

Adams, George. 1794. *Lectures on natural and experimental philosophy, considered in it's* [sic] *present state of improvement*, 5 vols. London.
Addison, Joseph. 1712. No. 413. *The Spectator*, June 24, 79–83.
Anonymous. 1747. *Verses, occasion'd by the sight of a chamera obscura*. London.

---

[86] Vestina [Graham] 1782, 63–64.
[87] For a discussion of these environments, see Otto 2011a, 144–159.
[88] Coleridge 1957–1990, Vol. 1 (Text), N. 1706.
[89] Coleridge 1830/1976, 29.

Armstrong, Nancy, and Leonard Tennenhouse. 2006. A mind for passion: Locke and Hutcheson on desire. In *Politics and the passions: 1500–1850*, ed. Victoria Kahn, Neil Saccamano, and Daniela Coli, 131–150. Princeton: Princeton University Press.

Beattie, James. 1783. *Dissertations moral and critical*, 2 vols. Dublin.

Burke, Edmund. 1790. *Reflections on the revolution in France, and on the proceedings of certain societies in London relative to that event*. London.

Campbell, Colin. 1987. *The romantic ethic and the spirit of modern consumerism*. Oxford: Blackwell.

Canning, George. 1801. New morality. In *Poetry of the Anti-Jacobin*, 4th ed, 233–256. London.

Coleridge, Samuel Taylor. 1848. *The poems of S. T. Coleridge*. London: William Pickering.

Coleridge, Samuel Taylor. 1797. Review of The Monk: A romance. By M.G. Lewis, Esq. M.P. *The Critical Review*, February 19, 194–200.

Coleridge, Samuel Taylor. 1800. The voice from the side of Etna; or, the mad monk. An ode in Mrs. Ratcliff's manner. *The Morning Post and Gazetteer*, October 13, 3.

Coleridge, Samuel Taylor. 1830/1976. *On the constitution of church and state*, ed. John Colmer. Princeton: Princeton University Press.

Coleridge, Samuel Taylor. 1895. *Letters of Samuel Taylor Coleridge*, 2 vols., ed. Ernest Hartley Coleridge. London: William Heinemann.

Coleridge, Samuel Taylor. 1932. *Unpublished letters of Samuel Taylor Coleridge*, 2 vols., ed. Earl Leslie Griggs. London: Constable.

Coleridge, Samuel Taylor. 1957–1990. *Notebooks of Samuel Taylor Coleridge*, 4 vols., ed. Kathleen Coburn. New York: Pantheon Books.

Coleridge, Samuel Taylor. 1983. *Biographia Literaria*, 2 vols., eds. J. Engell and W. Jackson Bate. *The collected works of Samuel Taylor Coleridge*, 16 vols., vol. 7. London/Princeton: Princeton University Press.

Cooper, Anthony Ashley, Earl of Shaftesbury. 1723. *Characteristicks of men, manners, opinions, times*, 3rd ed, 3 vols. London.

Crary, Jonathan. 1992. *Techniques of the observer: On vision and modernity in the nineteenth century*. Cambridge, MA: MIT Press.

Derrida, Jacques. 2008. *The Gift of Death & Literature in Secret*. Trans. D. Wills. Chicago/London: University of Chicago Press.

Descartes, René. 1985. *The Philosophical Writings of Descartes*, 2 vols. Trans. J. Cottingham, R. Stoothoff, and D. Murdoch. Cambridge: Cambridge University Press.

Gaston, Sean. 2010. The impossibility of sympathy. *Eighteenth Century: Theory & Interpretation* 51(1–2): 129–152.

Goodrich, Peter. 1995. *Oedipus Lex: Psychoanalysis, history, law*. Berkeley: University of California Press.

Home, Henry, Lord Kames. 1785. *Elements of criticism*, 6th ed, 2 vols. Edinburgh.

Howells, Coral Ann. 1995. *Love, mystery and misery: Feeling in gothic fiction*. London: Athlone.

Huckvale, David. 2010. *Touchstones of gothic horror: A film genealogy of eleven motifs and images*. Jefferson: McFarland & Co.

Hume, David. 1739–1740/1985. *A treatise of human nature*, ed. Ernest C. Mossner. London: Penguin.

Hutcheson, Francis. 1728. *An essay on the nature and conduct of the passions and affections*. Dublin.

Jones, Wendy. 1990. Stories of desire in The Monk. *ELH* 57: 129–150.

Krzywinska, Tanya. 2006. *Sex and the cinema*. London: Wallflower.

Lewis, Matthew Gregory. 1796/1973. *The Monk: a romance*, ed. Howard Anderson. London: New English Library.

Locke, John. 1690/1959. *An essay concerning human understanding*, 2 vols., ed. Alexander Campbell Fraser. New York: Dover.

Lyotard, Jean-François. 1997. Answering the Question: What Is Postmodernism. Trans. R. Durand. In *The Postmodern Condition: A Report on Knowledge*, 71–82. Trans. Geoff Bennington and Brian Massumi. Minneapolis: University of Minnesota Press.

Macdonald, David Lorne. 2000. *Monk Lewis: A critical biography*. Toronto: University of Toronto Press.
Mathias, Thomas. 1797. *The Pursuits of literature: A satirical poem in dialogue. With notes. Part the fourth and last*. London.
Mathias, Thomas. 1798. *The Pursuits of literature: A satirical poem in four dialogues. With notes*, 8th ed. London.
Mishra, Vijay. 1994. *The Gothic sublime*. Albany: State University of New York Press.
Mulman, Lisa Naomi. 1998. Sexuality on the surface: Catholicism and the erotic object in Lewis's *The Monk*. In *Making history: Textuality and the forms of eighteenth century culture*, ed. Greg Clingham, 98–110. Cranbury: Associated University Presses.
Newton, Isaac. 1718. *Opticks, or, a treatise of the reflections, refractions, inflections and colours of light*, 2nd ed. London.
Norton, Rictor. 2000. *Gothic readings: The first wave, 1764–1840*. London: Leicester University Press.
Otto, Peter. 2011a. *Multiplying worlds: Romanticism, modernity, and the emergence of virtual reality*. Oxford: Oxford University Press.
Otto, Peter. 2011b. Inside the imagination-machines of gothic fiction: Estrangement, transport, affect. In *Minds, bodies, machines, 1770–1930*, ed. Deirdre Coleman and Hilary Fraser, 19–38. Basingstoke: Palgrave Macmillan.
Paine, Thomas. 1791. *Rights of man: Being an answer to Mr. Burke's attack on the French revolution*, 8th ed. London.
R. R. 1797. Review of *The Monk. A romance*. By M. G. Lewis, Esq. M.P. *The European magazine, and London review*. February, 111–115.
Sedgwick, Eve Kosofsky. 1981. Character in the veil: Imagery of the surface in the Gothic novel. *PMLA* 96: 255–270.
Stafford, Barbara Maria, and Frances Terpak. 2001. *Devices of wonder: From the world in a box to images on a screen*. Los Angeles: Getty Research Institute for the History of Art & the Humanities.
Swann, Karen. 1985. Literary gentlemen and lovely ladies: The debate on the character of Christabel. *ELH* 52: 397–418.
Ussher, James. 1769. *Clio: or, a discourse on taste. Addressed to a young lady*, 2nd ed. London.
Vestina, Hebe [James Graham]. 1782. *Il Convito amoroso! Or, a serio-comico-philosophical lecture on the causes, nature, and effects of love and beauty*, 2nd ed. London.
Žižek, Slavoj. 1997. *The Plague of fantasies*. London: Verso.

CPSIA information can be obtained at www.ICGtesting.com
Printed in the USA
LVOW10*1915221213

366423LV00008B/63/P